RESEARCH AND PERSPECTIVES IN ALZHEIMER'S DISEASE

Fondation Ipsen

Editor

Yves Christen, Fondation Ipsen, Paris (France)

W0042296

Editorial Board

F. Gage A. Privat Y. Christen (Eds.)

Neuronal Grafting and Alzheimer's Disease

Springer-Verlag
Berlin Heidelberg New York
London Paris Tokyo Hong Kong

Gage, F. H., Dr.
Neurosciences, School of Medicine
M. 024 University of California
La Jolla, CA 92093, USA

Privat, A., Dr.
Laboratoire de Neurobiologie
Institut de Biologie
Boulevard Henri IV
F-34060 Montpellier

Christen, Yves, Dr.
Fondation IPSEN
pour la Recherche Therapeutique
30, rue Cambronne
F-75737 Paris Cedex

ISBN-13: 978-3-642-48371-4   e-ISBN-13: 978-3-642-48369-1
DOI: 10.1007/978-3-642-48369-1

© Springer-Verlag Berlin Heidelberg 1989
Softcover reprint of the hardcover 1st edition 1989

The use of general descriptive names, trade names, trade marks, etc. in this publication, even if the former are not especially identified, is not to be taken as a sign that such names, as understood by the Trade Marks and Merchandise Marks Act, may accordingly be used freely by anyone.

Product Liability: The publisher can give no guarantee for information about drug dosage and application thereof contained in this book. In every individual case the respective user must check its accuracy by consulting other pharmaceutical literature.

2127/3140/543210 − Printed on acid-free paper

Preface

This volume contains the proceedings of the third Colloque Medecine et Recherche organized by the Fondation Ipsen pour la Recherche Thérapeutique and devoted to Alzheimer's disease. After *Immunology and Alzheimer's disease* (Angers, September 14, 1987) and *Genetics and Alzheimer's disease* (Paris, March 25, 1988), this meeting was dedicated to an other up-to-date, and even futurist, topic: *Neuronal grafting and Alzheimer's disease*. It was held in Montpellier, an historical city, and under the patronage of one of the oldest School of Medicine of the world.

For the last few months, a large number of experimental works and even of clinical trials on neuronal grafting have been carried out. As far as clinical studies are concerned, results remain highly controversial. Most of them concern Parkinson's disease but for some scientists, neuronal grafting could be a cure for other neurodegenerative diseases, including Alzheimer's disease (and also Huntington's disease) but, at this point, there are only a few experimental data (perhaps not really related to the disease) on animal models and hypothesis. So the main aim of the Montpellier's meeting was not, indeed, to give a definitive answer to these questions.

The following Colloque Médecine et Recherche of this series was organized in Toulouse on March 24, 1989, and was dedicated to *the biological markers of Alzheimer's disease*.

Yves Christen
Vice-Président of the Fondation Ipsen pour la Recherche Thérapeutique

Acknowledgments: The editors wish to express their gratitude to Dean Claude Solassol for giving the patronage of the faculty of medicine of Montpellier, Mrs Mary Lynn Gage for her editorial assistance, Mrs Jacqueline Mervaillie for the organization of the meeting and Yves Agid and Jean-Didier Vincent for their collaboration as chairmen for the meeting.

Contents

Contributors

Abrous, N.
  INSERM U 259, Université Bordeaux II., Domaine de Carreire,
  rue Camille Saint-Saens, 33077 Bordeaux, France

Ayer-LeLievre, C.
  Department of Histology and Neurobiology, Karolinska Institute,
  P.O.Box 60400, 104 01 Stockholm, Sweden

P. Bickford-Wimer
  Department of Pharmacology (C236) University of Colorado,
  Health Sciences Center, 4200 East Ninth Avenue, Denver, CO 80262, USA

Björklund, A.
  Department of Medical Cell Research, Biskopsgatan 5, 223 62 Lund, Sweden

Brundin, P.
  Department of Medical Cell Research, University of Lund,
  Biskopsgatan 5, 22362 Lund, Sweden

Buzsáki, G.
  Department of Neurosciences M-024, University of California at San Diego,
  La Jolla, CA 92093, USA

Cassel, J.-C.
  Département de Neurophysiologie et de Biologie des Comportments, Centre de
  Neurochimie du C.N.R.S., 12 rue Goethe, 67000 Strasbourg, France

Choulli, K.
  INSERM U 259, Université Bordeaux II., Domaine de Carreire,
  rue Camille Saint-Saens, 33077 Bordeaux, France

Christen, Y,
  Fondation Ipsen, 30 rue Cambronne 7574 Paris Cedex 15

Collier, T. J.
  Department of Neurobiology and Anatomy, University of Rochester,
  School of Medicine, Rochester, NY 14642, USA

Cotman, C. W.
  Department of Psychobiology, University of California, Irvine, CA 92717, USA

Drian, M. J.
  INSERM U 249 et CNRS LP 8402, 34060 Montpellier, France

Dunnett, S. B.
  Department of Experimental Psychology, University of Cambridge,
  Downing Street, Cambridge CB2 3EB, UK

Dusart, I.
  INSERM U 161, 2 rue d'Alesia, Paris 75014, France

Ebendal, T.
  Department of Developmental Biology, Biomedical Center,
  Uppsala University, 751 23 Uppsala, Sweden

Eriksdotter-Nilsson, M.
  Department of Histology and Neurobiology, Karolinska Institute,
  P.O.Box 60400, 104 01 Stockholm, Sweden

Ernfors, P.
  Department of Medical Chemistry Laboratory of Molecular Neurobiology,
  Karolinska Institute, P.O.Box 60400, 104 01 Stockholm, Sweden

Fares, F.
  Rappaport Family Institute for Research in the Medical Sciences,
  Department of Pharmacology, Faculty of Medicine, Technion-Israel,
  Institute of Technology, Haifa, Israel

Favier, F.
  Laboratoire Cytométrie Flux INSERM, 34060 Montpellier, France

Fuentès, C.
  Laboratore de Neurophysiologie, U.S.T.L. Place E. Bataillon,
  34060 Montpellier Cedex 1, France

Gage, F. H.
  Department of Neurosciences M-024, University of California at San Diego,
  La Jolla, CA 92093, USA

Gansmuller, A.
  INSERM U 134, Laboratoire de Neurobiologie cellulaire,
  moléculaire et clinique, 47 Boulevard de l'Hôpital, Hôpital de la Salpétriére,
  75651 Paris Cedex 13, France

Gavish, M.
Rappaport Family Institute for Research in the Medical Sciences,
Department of Pharmacology, Faculty of Medicine, Technion-Israel,
Institute of Technology, Haifa, Israel

Geffard, M.
Institut de Biochimie Cellulaire et de Neurochimie, 33007 Bordeaux, France

Gerhardt, G.
Department of Pharmacology (C236), University of Colorado,
Health Sciences Center, 4200 Ninth Avenue, Denver, CO 80262, USA

Gill, T. J.
University of Pittsburgh, School of Medicine, Pittsburgh, PA 15261, USA

Gout, O.
INSERM U 134, Laboratoire de Neurobiologie cellulaire,
moléculaire et clinique, 47 Boulevard de l'Hôpital, Hôpital de la Salpétriére,
75651 Paris Cedex 13, France

Granholm, A.-C.
Karolinska Institute, P.O. Box 60400, 104 10 Stockholm, Sweden

Greenberger, V.
Center for Neurosciences, Weizmann Institute of Science, Rehovot, Israel

Greenfeld, Z.
The Melvin A. and Eleanor Ross Laboratory for Studies in Neural Birth
Defects, Department of Anatomy and Embryology, The Hebrew University-
Hadassah Medical School, Box 1172, 91010 Jerusalem, Israel

Gumpel, M.
INSERM U 134, Laboratoire de Neurobiologie cellulaire,
moléculaire et clinique, 47 Boulevard de l'Hôpital, Hôpital de la Salpétriére,
75651 Paris Cedex 13, France

Herman, J. P.
INSERM U 259, Université Bordeaux II., Domaine de Carreire,
rue Camille Saint-Saens, 33077 Bordeaux, France

Hoffer, B.
Department of Pharmacology (C236), University of Colorado,
Health Sciences Center, 4200 East Ninth Avenue, Denver, CO 80262, USA

Isacson, O.
Department of Anatomy, University of Cambridge, Downing Street,
Cambridge CB2 3DY, UK

Kelche, C.
Département de Neurophysiologie et de Biologie des Comportements,
Centre de Neurochimie du C.N.R.S., 12 rue Goethe, 67000 Strasbourg, France

König, N.
Laboratore de Neurophysiologie, U.S.T.L., INSERM U249 et CNRS LP 8402
Place E. Bataillon, 34060 Montpellier Cedex 1, France

Le Moal, M.
INSERM U 259, Université Bordeaux II., Domaine de Carreire,
rue Camille Saint-Saens, 33077 Bordeaux, France

Lund, R. D.
University of Pittsburgh, School of Medicine, Pittsburgh, PA 15261, USA

Mouton, P.
Department of Histology and Neurobiology, Karolinska Institute,
P.O.Box 60400, 104 01 Stockholm, Sweden

Nothias, F.
INSERM U 161, 2 rue d'Alesia, 75014 Paris, France

Olson, L.
Department of Histology and Neurobiology, Karolinska Institute,
P.O.Box 60400, 104 01 Stockholm, Sweden

Onteniente, B.
Centre de Biochimie, Université de Nice, 06034 Nice Cedex, France

Palmer, M.
Department of Pharmacology (C236), University of Colorado,
Health Sciences Center, 4200 East Ninth Avenue, Denver, CO 80262, USA

Persson, H.
Department of Medical Chemistry, Laboratory of Molecular Neurobiology,
Karolinska Institute, P.O.Box 60400, 104 01 Stockholm, Sweden

Peschanski, M.
INSERM U 161, 2 rue d'Alesia, 75014 Paris, France

Pick, C. G.
The Melvin A. and Eleanor Ross Laboratory for Studies in Neural Birth
Defects, Department of Anatomy and Embryology, The Hebrew University-
Hadassah Medical School, Box 1172, 91010 Jerusalem, Israel

Privat, A.
Laboratoire de Neurobiologie, Institut de Biologie, Boulevard Henri IV, 34060 Montpellier, France

Rajaofetra, N.
INSERM U 249 et CNRS LP 8402, 34060 Montpellier, France

Rao, K.
University of Pittsburgh, School of Medicine, Pittsburgh, PA 15261, USA

Richter-Levin, G.
Center for Neurosciences, Weizmann Institute of Science, Rehovot, Israel

Rogel-Fuchs, Y.
The Melvin A. and Eleanor Ross Laboratory for Studies in Neural Birth Defects, Department of Anatomy and Embryology, The Hebrew University-Hadassah Medical School, Box 1172, 91010 Jerusalem, Israel

Rose, G.
Department of Pharmacology (C236), University of Colorado, Health Sciences Center, 4200 East Ninth Avenue, Denver, CO 80262, USA

Sandillon, F.
INSERM U 249 et CNRS LP 8402, 34060 Montpellier, France

Segal, M.
Center for Neurosciences, Weizmann Institute of Science, Rehovot, Israel

Seiger, A.
Karolinska Institute, P.O.Box 60400, 104 01 Stockholm, Sweden

Shpiegelman, R.
Center for Neurosciences, Weizmann Institute of Science, Rehovot, Israel

Sladek Jr., J. R.
Department of Neurobiology and Anatomy, University of Rochester, School of Medicine, Rochester, NY, 14642, USA

Sotelo, C.
Laboratoire de Neuromorphologie, INSERM U 106, Hôpital de la Salpétrière, 47 Boulevard de l'Hôpital, 75651 Paris Cedex 13, France

Strömberg, I.
Department of Histology and Neurobiology, Karolinska Institute, P.O.Box 60400, 104 01 Stockholm, Sweden

Trombkal, D.
The Melvin A. and Eleanor Ross Laboratory for Studies in Neural Birth
Defects, Department of Anatomy and Embryology, The Hebrew University-
Hadassah Medical School, Box 1172, 91010 Jerusalem, Israel

Will, B.
Département de Neurophysiologie et de Biologie des Comportements,
Centre de Neurochimie du C.N.R.S., 12 rue Goethe, 67000 Strasbourg, France

Yanai, J.
The Melvin A. and Eleanor Ross Laboratory for Studies in Neural Birth
Defects, Department of Anatomy and Embryology, The Hebrew University-
Hadassah Medical School, Box 1172, 91010 Jerusalem, Israel

Neuronal Grafting and Alzheimer's Disease: Introduction

F. H. Gage, A. Privat, and *Y. Christen*

Grafting tissue to the brain as possible therapy for neurodegenerative disease has captured the imagination of the lay public and of many scientists. For the lay public this awakened interest has led to hopes for therapy and cure, very soon. For the enlightened neuroscientist, the awakened interest has led to dedicated research into basic issues that need to be resolved to determine if this potential therapy is credible. The identification of specific neurodegenerative diseases which may be amenable to grafting therapy — Parkinson's disease, Huntington's disease, and the focus of the present volume, Alzheimer's disease — has heightened the concern, and in many cases the immediacy of the expectations, of those most directly involved with these devastating diseases.

At the core of this potential therapy lies basic research which does not necessarily have application to a clinically relevant disease as its common goal. For many of the basic scientists who have provided the information that has led to the present level of excitement, intracerebral grafting is to be used for learning more about the neuronal plasticity of the adult nervous system, and as a technique to establish principles of neuronal development. To these ends, intracerebral grafting is established as a powerful tool in the arsenal of the neurobiologist.

En route to the establishment of grafting to the brain as a useful methodology, the basic biological requirements for survival of grafted tissue had to be identified. While many important issues have been recognized, more research is needed to clarify not only optimal survival but also optimal function of the grafts. Some of the issues that have been identified as crucial for graft survival are:
1. donor age,
2. host age,
3. availability of trophic and tropic factors,
4. immunology of the brain,
5. vascularization of the graft, and
6. target specificity.

For most donor tissues, the younger the cells the better the chance of survival. This is particularly true of neuronal tissue, which explains the focus of attention on fetal cells in intracerebral grafting. An equally important factor is the age of the host brain. While grafts will survive in the adult and aged brain (albeit less well in the aged brain) significantly better survival is observed in the young neonatal brain. Both the limitations of donor age and host age may be attributed in part to the availability of trophic and tropic factors which can address specifically or non-specifically the survival and growth of grafted tissue independently of many other issues. One issue that will be difficult to circumvent, however, is that of immunological rejection. While early reports advocated a privileged status for the brain, recent investigations have made it clear that the brain is not immunologically privileged and that successful grafting to the brain must take this barrier into consideration.

The last two issues cited above are common-sense issues. In order for a graft to survive and thrive in the brain it must be appropriately vascularized, and thereby receive essential nutriments from the environment, and yet preserve, where essential, the blood-brain barrier. Target specificity dictates that the chances of survival for a donor tissue are improved if the donor tissue is placed near or in the area of the brain that the grafted neurons normally innervate. While these survival issues are not exhaustive and much further information is needed to complete this list, they provide an outline for experimentalists to design and conduct grafting studies.

While the above factors can be used to influence the survival of the grafts, a major objective is to establish graft function in the host brain. To this end it is important to delineate what functions can be expected from a grafted tissue, since those expectations should influence the type of tissue and the animal model that is used. Intracerebral grafting often evokes the image of cells replacing exactly the anatomical, biochemical, and physiological locations and actions of lost or damaged cells in the brain. This is the most rigid of expectations and one that has not been achieved to date. With as small a variation as misplaced synapses to the spine head versus spine neck, dysfunction rather than restored function can be induced by grafted cells. At another level of expectation a graft may have some interactions with the host brain, sufficient to provide a partial replacement of missing connections. Still further from exact integration, a graft could act as an endogenous minipump secreting a missing transmitter in the vicinity of vacant receptors. This secretory function could be either regulated or nonregulated and still be expected to have some replacement value. Some grafted tissue may not secrete transmitter-specific information at all but rather secrete trophic or tropic factors which directly affect the survival and growth of damaged cells in the host brain. Finally, grafts can function as different types of bridges. A bridge that acts as a conduit between two disconnected portions of the brain but does not contribute to the connection on either side can be viewed as a passive bridge, while a bridge that functions as a conduit and also contributes neuronal and axonal connections to either side can be viewed as an active bridge. With this variety of expectations for graft function, it becomes clear that even the very notion that neuronal tissue is a requirement for intracerebral grafting becomes an unneeded limitation.

While the principles and issues related to the optimal survival and function of grafts to the brain continue to be elucidated, and the utility of intracerebral grafting becomes firmly established as a powerful tool in neurobiology, the application of this technique to neurological disease is not yet clear. Advances along these lines will require once again a definition of expectations. If a cure for the disease is sought, then neuronal grafting does not appear to be a likely approach. Before grafting could even be considered, the cause of the neurodegenerative disease would need to be discovered. If, however, providing therapy for some of the symptoms of the disease is sought, then intracerebral grafting may prove more realistic. However, two related hurdles inhibit advances in this latter approach for Alzheimer's disease. The first is a clear understanding of the underlying pathology which is responsible for the symptomology for which therapy is sought. Without this basic knowledge, the experimentalist working in the field of intracerebral grafting does not know "what cells to put in which part of the brain." Hand in hand with this need for an understanding of the pathophysiology of the disease is the need for experimental animal models which mimic the pathological characteristics of the disease, and in a correlated way, the cardinal functional symptoms.

To a large extent Parkinson's disease has met these two requirements of identified pathophysiology and analogous animal models. As a result the application of intracerebral grafting in Parkinson's disease has progressed to initial clinical trials. The next step − moving the clinical trials from the realm of experimental human research to standardized therapeutic surgery will undoubtedly take longer than expected, since new, previously unaddressed issues must be dealt with that involve science, clinical medicine, and society's willingness to understand and accept what science and medicine have to offer.

The wealth of basic information presented in this volume testifies to the progress that has been achieved in intracerebral grafting and to the utility of intracerebral grafting as a tool for the understanding of brain development, adult neuronal plasticity, and age-related pathology. An answer to the question of whether neuronal grafting will be useful as a therapy for Alzheimer's disease must wait for a better understanding of the disease and the identification of animal models that can be used to test potential therapies. Meanwhile, progress in optimizing graft survival and function will continue, and the tool of intracerebral grafting may, in the future, be used to address the pathophysiology of Alzheimer's disease.

Nigral and Striatal Grafts to the Lesioned Neostriatum: Models of Graft Function

A. Björklund, O. Isacson, P. Brundin, and *S. B. Dunnett*

Summary

The ability of intracerebrally implanted grafts of neural tissues to promote functional recovery in brain damaged recipient animals has raised the question of how such implants exert their functional effects. Nonspecific, diffuse release of active compounds may be sufficient to restore defective neurotransmission in a denervated brain region, for example, or to provide trophic support for the survival and regeneration of damaged host neurons. The positive therapeutic effects of adrenal medullary grafts, recently reported in patients with Parkinson's disease, are likely to reflect such nonspecific hormonal or neurohumoral mechanisms. Morphological and electrophysiological studies, on the other hand, have shown that grafted fetal neurons can establish extensive efferent synaptic connections with previously denervated or neuron-depleted host brain regions and become at least partially integrated into the host neuronal circuitry. In the damaged nigrostriatal system, grafts of fetal nigral or striatal neurons can restore normal synaptic transmitter release and can also participate in a partial reconstruction of functional neural circuits in the host brain. This indicates that the potential of intracerebral grafts to induce or improve behavioral recovery in brain-damaged recipients rests on the multitude of trophic, neurohumoral and synaptic mechanisms that may allow the implanted tissue to promote host brain function and repair.

Introduction

The classical perspective, derived above all from Cajal's work (R. Y. Cajal 1928), that all regeneration in the central nervous system of adult mammals is abortive, has been increasingly challenged over the past two decades, in particular following the observations of reactive synaptogenesis in deafferented brain regions (Raisman 1969; Lynch et al. 1972) and of regeneration of fine unmyelinated axons, as described for the central monoamine systems following axotomy (Katzman et al. 1971) and following implantation of Schwann-cell containing conduits, such as iris and sciatic nerve grafts (Svengaard et al. 1975; Aguayo et al. 1981). These observations were soon followed by the demonstration that embryonic CNS neurons could undergo even more dramatic growth when transplanted to damaged adult CNS targets (see Björklund and Stenevi 1984 for review), opening up the prospect of using transplantation techniques to reconstruct damaged circuits in experimental and potentially even clinical contexts. Over the past few years the potential for using neural grafting techniques to counteract age-related functional deficits has been extensively explored also in animal models of aging and dementia (Gash et al. 1985; Gage and Björklund 1986).

The most extensive work on the functional consequences of intracerebral transplantation techniques has so far been conducted in animals with damage to basal ganglia circuitries. Since several reviews already exist on the nature and extent of behavioral changes observed following intracerebral transplantation (Dunnett et al. 1983c, d, 1985, 1986b; Freed 1983; Björklund et al. 1987), the present chapter will focus instead on the levels of reorganization and models of graft functions that may account for those changes.

The Animal Lesion Model

Model Parkinsonism

Intracerebral injection of the neurotoxin 6-hydroxydopamine (6-OHDA) produces a selective destruction of catecholamine neurons. In particular, injection of 6-OHDA into the nigrostriatal pathway results in a depletion of striatal dopamine and a behavioral syndrome that provides a widely used model of human Parkinsonism (Schultz 1982; Marshall and Teitelbaum 1977; Stricker and Zigmond 1976). Bilateral injections produce profound akinesia, accompanied by aphagia and adipsia, such that the animals require intragastric feeding to be kept alive (Ungerstedt 1971c; Zigmond and Stricker 1972). By contrast, animals with unilateral lesions do not show these regulatory impairments and maintain normal health, but manifest marked motor and sensorimotor asymmetries characterized by spontaneous and amphetamine-induced turning ("rotation") to the side ipsilateral to the lesion (Ungerstedt 1971a), contralateral apomorphine-induced rotation (which is believed to be attributable to the development of receptor supersensitivity in the deafferented striatum; Ungerstedt 1971b), and contralateral sensory neglect (Marshall et al. 1974).

Model Chorea

Intrastriatal injection of kainic acid, ibotenic acid, or other excitotoxic amino acids produces a selective destruction of striatal neurons while sparing the cortical afferent and efferent fibres running in the internal capsule (Coyle and Schwarcz 1976; Schwarcz and Coyle 1977). When made bilaterally such lesions produce a behavioral syndrome that has been considered to provide a good model of Huntington's chorea, both in terms of the patterns of neuroanatomical degeneration (Coyle and Schwarcz 1976), and in that the lesioned rats are hyperkinetic (Mason and Fibiger 1978; Dunnett and Iversen 1981) and display "cognitive" impairments on tasks sensitive to prefrontal damage, such as delayed alternation (Divac et al. 1978), spatial reversal (Dunnett and Iversen 1981), and operant DRL learning (Dunnett and Iversen 1982).

Functional Recovery with Intrastriatal Grafts

Transplantation Procedures

Several alternative procedures have been used to transplant embryonic neurons to the denervated or deafferented neostriatum. Pieces of embryonic tissue can be inserted either into the lateral ventricles, adjacent to the medial surface of the caudate-putamen (Freed et al. 1980; Perlow et al. 1979), or into preformed aspirative cavities exposing the dorsal or lateral surfaces of the nucleus (Björklund and Stenevi 1979; Björklund et al. 1980a; Stenevi et al. 1980). Alternatively, dissociated suspensions of cells can be stereotactically injected directly into the host striatal parenchyma (Björklund et al. 1980b, 1983a; Schmidt et al. 1981). Each procedure is relatively routine and can be expected to yield an 85% or higher rate of graft survival.

Nigral Grafts to the 6-OHDA-Lesioned Striatum

There is now extensive evidence that intrastriatal grafts of embryonic nigral tissue rich in dopamine cells can substantially or completely ameliorate many of the motor and sensorimotor asymmetries of rats with unilateral 6-OHDA lesions, and also increase the spontaneous locomotor activity levels of akinetic rats with bilateral lesions (see Dunnett et al. 1983c, d, 1985, 1986; Freed 1983; Björklund et al. 1987 for reviews). By contrast, other behavioral sequelae of bilateral lesions remain uninfluenced by the presence of the grafts. This appears to be the case for skilled paw use (Dunnett et al. 1987), and hoarding behavior (Herman et al. 1986), as well as the aphagia and adipsia which is seen in bilaterally lesioned animals (Björklund et al. 1980a; Dunnett et al. 1981c, 1983b).

Striatal Grafts to the Ibotenic Acid-Lesioned Striatum

Grafts of embryonic striatal tissue have been shown to reverse the locomotor hyperactivity of lesioned rats (Deckel et al. 1983; Isacson et al. 1984, 1986; Sanberg et al. 1987) and to ameliorate the lesion-induced deficits in spatial alternation learning (Isacson et al. 1986; Deckel et al. 1986) and skilled paw use (Dunnett et al. 1988b). Biochemically, there is a significant recovery in both GABA and acetylcholine synthetic enzymes (Isacson et al. 1985) and in glucose metabolic rate in striatal output structures (Isacson et al. 1984).

Levels of Reconstruction of Lesioned Circuitries

Several different models of anatomical reorganization in the host brain might be proposed to account for the observed behavioral recovery, reflecting different mechanisms of action and different levels of reconstruction within the damaged host circuitries.

1. *Nonspecific consequences of transplantation.* Transplantation surgery itself could provoke changes in the host brain, but of a nature unrelated to the nature of the graft tissue or its location. Such changes may be related, e.g., to opening of the blood-brain barrier, invasion of macrophages or glia, or increased production of trophic factors.
2. *"Pharmacological" action.* The grafts may secrete biologically active molecules (e.g., transmitters or trophic factors) into host parenchyma, cerebrospinal fluid, or circulation that exert their effects by diffuse action.
3. *"Tonic" reinnervation.* The grafts may establish a partial or complete reinnervation of denervated host targets which restores tonic activation of, but is unregulated by dynamic changes in, the host brain.
4. *Reciprocal reinnervation.* The grafts may additionally become subject to partial control from the host, either by diffuse neurohormonal influences or by partial collateral reinnervation from neurons of the host brain.
5. *Full reconstruction.* The grafts reciprocally establish connections to and from the host brain appropriate to the populations of cells damaged by the host lesion and replaced by transplantation from the donor. This would allow the grafted neuronal elements to become at least partly incorporated into the host neuronal circuitry.

These levels of reconstruction are not necessarily discrete or mutually exclusive. Patterns of graft-induced reinnervation may only be partial, and grafts may develop abnormal as well as normal connections. Furthermore, it seems quite possible that several types of effects, e.g., nonspecific effects of the transplantation surgery, "pharmacological" types of effects, and connectivity-mediated functions may interact and collaborate in the promotion of functional recovery.

Models of Nigral Graft Function

Nonspecific Models

Nonspecific effects (level 1, above) may account for some aspects of the functional recovery seen following dopamine-rich grafts in 6-OHDA-lesioned rats. For example, the trauma of graft surgery may perhaps produce a decline in rotational and sensorimotor asymmetry by means of a generalized decrease in activity or attention to both sides. Such lesion-induced changes would, however, not explain other graft-induced functional effects, such as the increased locomotor activity in akinetic rats with bilateral lesions, and the observation that in nigra-grafted animals sensorimotor symmetry is restored by an increase of contralateral attention. In 6-OHDA-lesioned rats nigral grafts transplanted to nonstriatal targets have no effect on the lesion syndrome (Dunnett et al. 1983a); neither do control grafts of nondopaminergic embryonic neural tissue transplanted to otherwise effective striatal sites (sciatic nerve, Freed et al. 1980; striatum and raphe, Dunnett et al. 1988a).

A more powerful model, which overlaps levels 1 and 2, is to suggest that embryonic tissue is rich in growth-promoting substances and trophic factors that can promote regenerative changes within the host brain itself (Manthorpe et al. 1983; Nieto-Sampedro et al. 1982; Kesslak et al. 1986; Bohn et al. 1987). Long-term graft survival might not be necessary then since the grafts would have no further functional contribution to make once asymptotic recovery is reached. However, in two studies (Björklund et al. 1980a; Dunnett et al. 1988a) the nigral grafts were removed once complete recovery of the amphetamine rotation response had taken place, and the initial high levels of asymmetry were immediately restored. Consequently, in the 6-OHDA-lesion model the continuous presence of the graft appears to be necessary in order to maintain the stable amelioration of symptoms. The observations of Bohn et al. (1987), working with MPTP-lesioned mice, suggest, however, that the situation may be different for grafts of adrenal medullary tissue.

Pharmacological Models

It seems likely that a level 2 (pharmacological) model is sufficient to explain the functional effects of some types of catecholamine-rich grafts. Thus, adrenal medullary tissue grafted to the ipsilateral lateral ventricle of rats with unilateral 6-OHDA lesions produces a partial reduction of apomorphine-induced rotation (Freed et al. 1981). This is accompanied by a halo of diffuse catecholamines in the host parenchyma (Freed et al. 1983) close to the graft. Similar, but transient, effects have been reported after grafting into the striatal parenchyma (Strömberg et al. 1985; Bing et al. 1988). Since peripheral injections of L-dopa or apomorphine can reverse many of the motor impairments in Parkinsonian patients or experimental animals with extensive dopaminergic deafferentation (Schultz 1982; Ljungberg and Ungerstedt 1976; Marshall and Ungerstedt 1976; Marshall and Gotthelf 1979), it seems possible that the recovery seen following adrenal grafts (and maybe also intraventricular nigral grafts; Freed et al. 1980), may act via a similar pharmacological, diffuse release process. This idea of a "biological minipump" has indeed been the primary rationale behind the use of adrenal-chromaffin cells in clinical trials.

A Tonic Model

Although a pharmacological model may sufficiently account for the functional effects of chromaffin tissue grafts in the 6-OHDA striatum, this may be insufficient for the effects of embryonic central catecholamine-rich grafts. Whereas the effects of adrenal medullary grafts are seen almost immediately following transplantation, and subsequently decline to a stable low level (Strömberg et al. 1985), grafts of embryonic nigral tissue take some weeks to become established before functional effects are detectable (Björklund and Stenevi 1979; Björklund et al. 1980a; Dunnett et al. 1981a, 1983a). Compensation of lesion-induced asymmetries develops progressively over a period of 2–6 months, in parallel with the establishment of an extensive fibre reinnervation of the host brain (Björklund et al. 1980a, 1983b). Moreover, the extent of compensation in different animals has been seen to correlate better with the extent of fibre reinnervation than with the size of the transplants or the number of dopaminergic neurons within the grafts (Björklund et al. 1980a). This suggests that reinnervation, rather than diffuse release of the deficient transmitter, may be important for functional recovery in the nigral-grafted animals. This notion is further supported by the observation that the graft-derived reinnervation establishes ultrastructurally normal synaptic contacts with appropriate postsynaptic targets in the deafferented host striatum (Freund et al. 1985; Mahalik et al. 1985).

Our current model of nigral graft function (Björklund et al. 1981; Dunnett et al. 1983d) is based on Stricker and Zigmond's (1976) model of normal nigrostriatal function (see Fig. 1). Their studies of the relationship between compensatory biochemical changes and functional recovery following nigrostriatal lesions led to the proposal that the nigrostriatal dopaminergic system regulates behavioral responsiveness through a modulation of an inhibitory striatal control of motor systems. Removal of this regulatory control by a 6-OHDA lesion results in an inhibition of behavioral responses (see Fig. 1b). Graft-derived reinnervation of the deafferented neostriatum, therefore, can be interpreted in terms of a reactivation of an inhibited but otherwise intact striatal circuitry (see Fig. 1c). This provides a

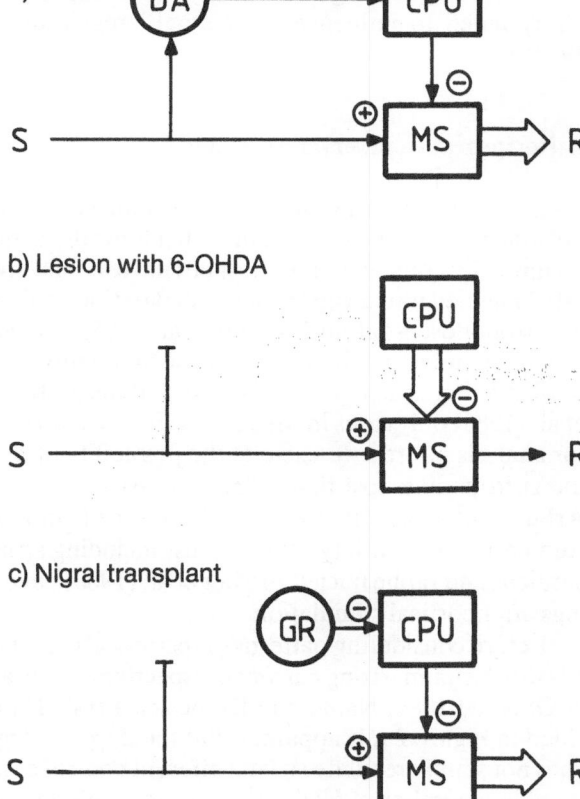

Fig. 1a–c. A tonic model of nigral graft reconstruction of damaged dopaminergic inputs to the neostriatum. **a** Organization of the intact system. **b** Effects of 6-OHDA lesion of intrinsic dopamine neurons of the nigrostriatal projection. **c** Graft-derived reinnervation of the deafferented neostriatum. *DA,* dopamine neurons of the substantia nigra pars compacta; *CPU,* caudate-putamen nucleus of the neostriatum; *GR,* nigral grafts; *MS,* motor systems; *R,* response outputs of the system; *S,* sensory stimulus inputs to the system. Synaptic inputs are inhibitory (−) or excitatory (+). (Redrawn from Björklund et al. 1981)

tonic dopamine release model in which the graft provides a diffuse functional activation of an inhibited system but is not in itself dependent upon dynamic influences from the host.

This model can account for the fact that functional compensation can be
achieved by nigral grafts in spite of their ectopic location. A tonic release model
is also sufficient to explain the graft-induced normalization of striatal indices of
dopamine turnover and release (Schmidt et al. 1982, 1983; Zetterström et al. 1986;
Strecker et al. 1987).

However, several problems remain. First, little attention has been given to the
fact that intact dopamine neurons themselves are responsive to sensory stimulation (Chiodo et al. 1980; Nieoullon et al. 1977), which may suggest that normal
dopaminergic function involves a dynamic rather than simply a tonic regulation of
striatal circuitries. Second, very little information is available about patterns of
innervation established from host to graft. Third, the failure to restore normal
food and water intake in aphagic and adipsic rats has been interpreted in terms of
possible functional heterogeneity of the neostriatum, implying that some critical
area for these functions remains non-reinnervated. Whereas the principle of
topographic heterogeneity of striatal terminal fields has subsequently been well
demonstrated (Dunnett et al. 1981a, b, 1983c), such functional heterogeneity
appears insufficient to account for continuing aphagia and adipsia, since a series of
studies placing nigral grafts in many different potential striatal sites have all consistently failed to influence the animals' regulatory impairments (Dunnett et al.
1983b).

A Reciprocal Reinnervation Model

Recent studies have provided the first indications that the nigral grafts do not simply innervate the host striatum with a tonically activating input, but may become
reciprocally connected into host circuitries. One study of retrograde transport of
HRP injected into the grafts has failed to find evidence for any reciprocal innervation from host striatum (Freund et al. 1985). However, this may be a false negative. Dopamine cell bodies are most frequently seen to cluster close to the graft
border rather than in the central core of the graft, whereas in the study of Freund
et al. (1985) only cases in which the HRP was restricted to the center were retained
for analysis in order to exclude the possibility of diffusion of the HRP tracer into
the surrounding host tissue. By contrast, using electrophysiological techniques,
Arbuthnott et al. (1985) observed that cells in nigral grafts were responsive to
stimulation in a variety of host sites, including striatum, prefrontal cortex, locus
ceruleus and raphe nuclei. Fischer et al. (1988) have recently reported similar findings after cortical stimulation.

Before considering patterns of reciprocal graft-host innervation in more detail,
it is worth summarizing current perspectives on basal ganglia circuitry (see reviews
in Graybiel 1984; Nauta and Domesick 1984). From the simplified scheme provided in Figure 2 it is apparent that the dopaminergic innervation of the striatum
does not simply regulate striatal efferent control of movement, but rather is implicated in a number of feedback loops regulating both pyramidal and extrapyrami-

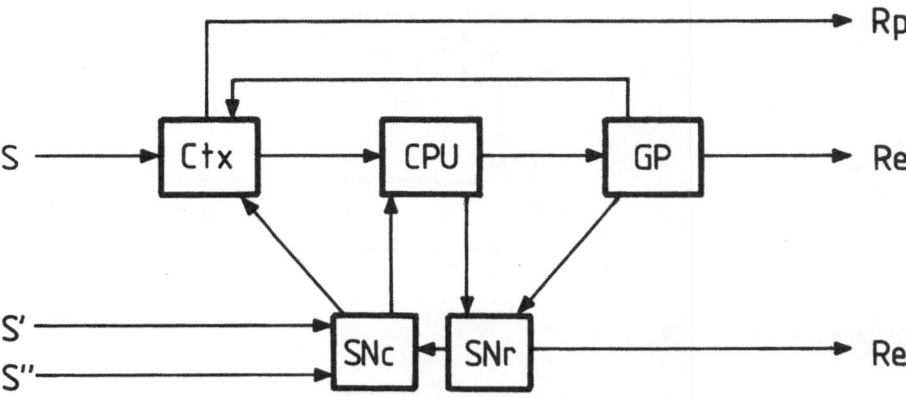

Fig. 2. Simplified scheme of the basic neuroanatomical organization of basal ganglia circuitry. *CPU*, caudate-putamen nucleus; *Ctx*, neocortex; *GP*, globus pallidus; *SN*, substantia nigra pars compacta *(c)* and reticulata *(r); S, S', S"*, inputs to the system at the level of the cortex and nigra; *R*, convergence on response outputs of the system via pyramidal *(p)* and extrapyramidal *(e)* motor systems. Note that thalamic relays, e.g., between *GP* and *Ctx*, have been omitted

dal motor systems (Rp and Re, respectively). Moreover, nigral dopamine neurons are themselves influenced by inputs (S', S") from several structures including the amygdala, lateral hypothalamus, and brain-stem nuclei.

Figure 3 suggests an alternative reciprocal innervation model of nigral graft function which incorporates the above observations on the normal anatomical organization and the physiological observations on host-graft connections, and which may overcome problems present in the simple tonic model. 6-OHDA lesion of the dopamine neurons of the substantia nigra pars compacta (SNc) disinhibits the striatal inhibition of pars reticulata (SNr) and pallidal (GP) control of both pyramidal and extrapyramidal motor systems (Fig. 3b). Additionally, the loss of the nigral dopamine neurons breaks both the striatonigral feedback regulation of basal ganglia circuitry and the influences of other afferent systems.

According to this model the nigral grafts provide a partial reconstruction of this circuitry (Fig. 3c). The grafts themselves comprise tissue taken from the whole ventral mesencephalon, including both pars reticulata and compacta of the substantia nigra. The electrophysiological recording studies (Arbuthnott et al. 1985) have indicated that, in addition to both dopaminergic and nondopaminergic innervation of the host striatum by the graft, afferents to the graft become established from the host striatum and neocortex and also from a number of brain-stem sites, including the locus coeruleus and raphe nuclei. Whether the "nigro"-striato-"nigral" loop is reestablished, as indicated in Figure 3c, is speculative, although rich interconnections between different types of neurons in the graft have been described in ultrastructural studies (Freund et al. 1985; Bolam et al. 1987).

The reinstatement by the grafts of a normal dopaminergic inhibition of striatal regulation of the initiation and control of motor responses proceeds in a manner similar to that proposed in the tonic model (Fig. 1), but in this case it is partially

a) Intact

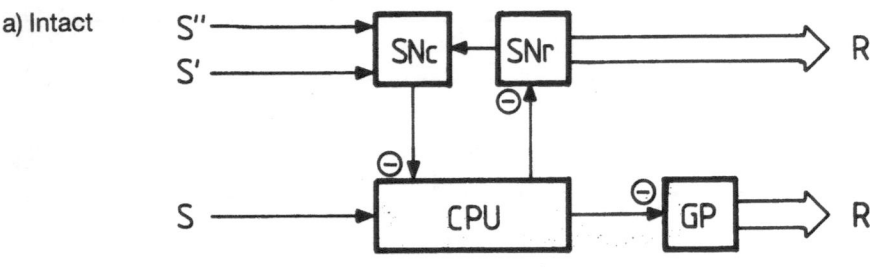

b) Lesion with 6-OHDA

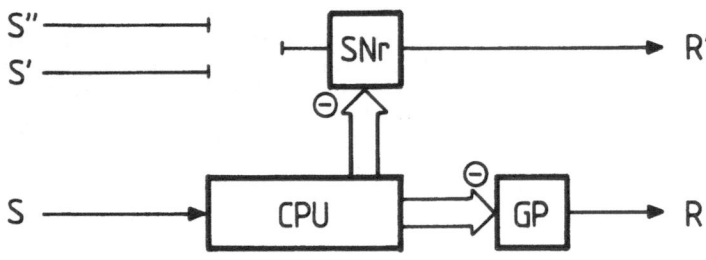

c) Nigral transplant

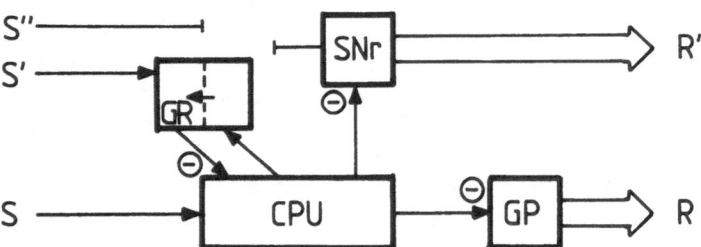

Fig. 3a–c. A reciprocal reinnervation model of nigral-graft reconstruction of damaged dopaminergic inputs to the neostriatum. **a** Organization of the intact system. **b** Effects of 6-OHDA lesion of intrinsic dopamine neurons of the nigrostriatal projection. **c** Graft-derived reinnervation of the deafferented neostriatum. Abbreviations as in Figs. 1 and 2

regulated by influences from both the basal ganglia, the brain-stem, and the frontal neocortex of the host. This may explain why complete recovery is seen in some classes of behavior, such as locomotor activity, motor bias, sensory attention, and conditioned turning (Dunnett et al. 1986b), whereas other types of dopamine-regulated behaviors remain uninfluenced, such as hoarding behavior (Herman et al. 1986), regulation of food and water intake, and skilled paw use (Dunnett et al. 1987). It has previously been shown that regulatory impairments produced by disruption of ascending dopamine systems are mimicked by neurotoxic lesions of the lateral hypothalamus (Winn et al. 1984). Winn et al. (1984) proposed that both types of lesion disconnect hypothalamic information about regulatory needs from striatal selection and initiation of the behaviors involved in seeking out and consuming food and water. Such motivational influences over the striatum, relayed via the substantia nigra (S" in Fig. 3c), may not become reestablished with the grafted nigral cells, and so the regulatory impairments remain unameliorated.

The simplest differentiation that can be proposed to distinguish between recovered and nonrecovered components of the nigrostriatal syndrome in the nigra-grafted animals might be between motor and activational impairments, on the one hand, versus motivational impairments, on the other. It might be argued that the difference between behaviors that are and are not influenced by nigral transplants is not those between reestablished and damaged neural circuits, but simply reflects the functional complexity of motivational as opposed to activational behaviors. This, however, seems unlikely. Fray et al. (1983) showed that grafted rats would lever-press to obtain stimulation by electrodes implanted in nigral but not control (cortical) grafts. Self-stimulation provides one of the classic and most direct methods of demonstrating motivational learning, and can be maintained by grafted rats when the relevant information is provided extrinsically to the grafted dopamine neurons. Thus, the failure of similar grafts to maintain normal food and water intake suggests that the grafted neurons do not receive the motivationally relevant information by which they normally influence the striatum. It seems unlikely, therefore, that the limited effects of the grafts on regulatory impairments are due to any essential incapacity of the reformed dopaminergic innervation of the striatum itself. This conclusion finds support in the recent study by Schwarz and Freed (1986) reporting that intrastriatal nigral grafts can prevent the development of aphagia and adipsia if the transplantation is made during the neonatal period.

In conclusion, we propose that the best model of dopamine graft function is provided by a level 4 model, involving the reformation of reciprocal influences between graft and host brain, but falling short of the full anatomical reconstruction necessary to achieve complete compensation of the behavioral syndrome. While it seems clear that host cells can reestablish connections with the intrastriatal nigral grafts, available evidence indicates that only some of the normal afferents to the substantia nigra are able to reinnervate the nigral graft in its ectopic location. It seems plausible, however, that greater reinnervation could become established if the grafted cells were replaced into the denervated host mesencephalon. Implantation of nigral cells into the substantia nigra region, however, produces no functional compensation on any of the behaviors that have been investigated (Dunnett et al. 1983a), in part at least because the grafted neurons appear to be unable to regenerate dopamine fibres over the distance necessary to reach the denervated striatum. One way to overcome this limitation may be to use additional peripheral nerve bridges to sustain axonal growth over the required distances (Aguayo et al. 1984), and preliminary studies of this model suggest that in some cases the sciatic nerve bridge can reconstruct some minimal functional connections, as measured by the compensation of amphetamine-induced rotation (Gage et al. 1985).

Models of Striatal Graft Function

The functional recovery seen following striatal grafts to the excitotoxin-lesioned striatum provides an even greater challenge to low-level nonspecific, pharmacological and tonic models. Unlike the motor impairments seen following 6-OHDA lesions, it seems unlikely that the deficits in complex, cortex-dependent behaviors, such as delayed alternation learning, induced by striatal lesions can be attributed exclusively to simple activational impairments. The neostriatum represents a major telencephalic relay which receives major cortical and thalamic inputs, and exerts control over the brain's sensorimotor functions via two major efferent outputs, the striopallidal and the strionigral pathways. On a simple level, large striatal lesions remove a major inhibitory (largely GABA-ergic) control of the globus pallidus and the substantia nigra pars reticulata, which in turn results in a disinhibition of subcortical motor driving systems, such as those in the subthalamic and mesencephalic locomotor regions. In the rat this disinhibition can most easily be measured in terms of locomotor hyperactivity as well as in terms of an increased glucose utilization in primary and secondary striatal target structures, including globus pallidus, substantia nigra, subthalamic nucleus, and the deep superior colliculus. Intrastriatal striatal grafts have been shown to counteract this locomotor hyperactivity in rats with unilateral or bilateral neostriatal excitotoxic lesions (Deckel et al. 1983, 1986; Isacson et al. 1984, 1986). This effect was accompanied by a normalization of the metabolic hyperactivity in several striatal output structures (Isacson et al. 1984). Sanberg and Henault (1986) have subsequently reported that the striatal grafts are effective only when they are located within the lesioned striatum, and not when placed in the adjacent lateral ventricle.

Striatal excitotoxic lesions also disrupt performance in more complex tasks, such as skilled paw-reaching and delayed alternation learning. Striatal grafts are also effective in ameliorating these types of behavioral impairments (Isacson et al. 1986; Deckel et al. 1986; Dunnett et al. 1988b). In the study by Isacson et al. (1986), the grafts were placed either into the head of the caudate-putamen or into the region of the globus pallidus. The neostriatal placement reduced both the hyperactivity and the delayed alternation impairments, whereas the pallidal placements were generally less effective.

The findings that neural implants can restore complex conditioned behaviors and skilled motor tasks after destruction of a major telencephalic cortical relay structure raise a number of interesting questions as to the mechanism(s) of action of the striatal grafts. In previous studies of dopaminergic and cholinergic neurons, grafted into the nigrostriatal system or the septohippocampal or basal forebrain-cortical systems, functional recovery in sensorimotoric or maze-learning tasks has been obtained by *ectopic* placement of the graft, i.e., within or in direct contact with the denervated striatal or cortical targets rather than in their normal location in the substantia nigra or the basal forebrain. In the striatal model *homotopic* graft placement (i.e., into the site of the lesion) appears to produce significantly better functional effects than ectopic placements, i.e., into globus pallidus or the adjacent lateral ventricle.

Recent neuroanatomical studies suggest that the intrastriatal striatal grafts can establish extensive reciprocal connections with the host brain (Pritzel et al. 1986;

Wictorin et al. 1988, 1989a; Wictorin and Björklund 1989). The striatal grafts have been shown to project to appropriate targets (such as the globus pallidus and in some cases probably also the substantia nigra) and to receive inputs from appropriate populations of host neurons, including the frontal cortex, amygdala, thalamus, substantia nigra, and the dorsal raphe. Electrophysiological studies both in vivo (Wilson et al. 1987) and in slices in vitro (Rutherford et al. 1987) have demonstrated functional connections between the host frontal cortex and neurons within the striatal grafts, and there is ultrastructural evidence for the formation of synaptic connections onto grafted neurons from both the ingrowing dopaminergic host nigral neurons (Clarke et al. 1988) and from the ingrowing host cortical neurons (Wictorin et al. 1989b). The demonstration of a rich dopaminergic input to the intrastriatal striatal grafts (Pritzel et al. 1986; Clarke et al. 1988) suggests that the host nigrostriatal dopamine pathway may play a primary role in the functional activation of the grafted striata. This is supported by a recent series of experiments showing that striatal grafts can compensate for the amphetamine- and apomorphine-induced turning response in rats with unilateral ibotenic acid lesions (Dunnett et al. 1988b).

The increased glucose utilization seen in the striatal output structures in rats with striatal excitotoxic lesions (Isacson et al. 1984) indicates that the caudate-putamen normally exerts an inhibitory tonic control on their functional activity. This suggests that it may be possible for the striatal grafts to normalize the metabolic and motoric hyperactivity in the ibotenic acid-lesioned rats by a tonic and relatively nonspecific GABA release. Although, in principle, such release could be of a diffuse neurohumoral nature, there is some evidence that the striatal grafts can reinstate GABA-ergic neurotransmission within the host globus pallidus. First, Wictorin et al. (1989a) have observed, with anterograde and retrograde neuroanatomical tracing techniques, extensive fibre outgrowth from the graft into the host globus pallidus (and perhaps also the substantia nigra, pars reticulata). Second, Isacson et al. (1985) reported a significant graft-induced recovery of glutamic acid decarboxylase activity in the host globus pallidus. And, finally, Sirinathsinghji et al. (1988), using the push-pull perfusion technique, have observed an approximately 30% recovery of GABA release in the globus pallidus region in ibotenic acid-lesioned rats with striatal grafts. Interestingly, in the latter study, the grafts were also seen to reinstate the responsiveness of pallidal GABA release to amphetamine, which is consistent with the idea that the grafted GABA-ergic neurons are under direct dopaminergic control.

Conclusion

Studies of neural grafts in the nigrostriatal system generally support the view that neural grafts can influence host brain function by different types of mechanisms and at different levels of complexity. In many experimental model systems the observed functional effects can most probably be explained in terms of diffuse release of deficient hormones or transmitters, tonic reafferentation of previously denervated areas of the host brain, or acute trophic influences over recovery phenomena in the damaged host. A minimal integration of the grafted neurons into the host neuronal circuitry is, however, likely to be necessary for certain types of graft-induced functional recovery to occur. The potential of intracerebral grafts to induce or improve behavioral recovery in brain-damaged recipients rests on the multitude of trophic, neurohumoral, and synaptic mechanisms that may allow the implanted tissue to influence host brain function or repair mechanisms.

References

Aguayo AJ, David S, Bray GM (1981) Influences of the glial environment on the elongation of axons after injury: transplantation studies in adult rodents. J Exp Biol 95:231–240

Aguayo AJ, Björklund A, Stenevi U, Carlstedt T (1984) Fetal mesencephalic neurons survive and extend long axons across peripheral nervous system grafts inserted into the adult rat striatum. Neurosci Lett 45:53–58

Arbuthnott G, Dunnett SB, MacLeod N (1985) Electrophysiological recording from nigral transplants in the rat. Neurosci Lett 57:205–210

Bing G, Notter MFD, Hansen JT, Gash DM (1988) Comparison of adrenal medullary, carotid body and PC12 cell grafts in 6-OHDA lesioned rats. Brain Res Bull 20:399–406

Björklund A, Stenevi U (1979) Reconstruction of the nigrostriatal dopamine pathway by intracerebral nigral transplants. Brain Res 177:555–560

Björklund A, Stenevi U (1984) Intracerebral neural implants: neuronal replacement and reconstruction of damaged circuitries. Annu Rev Neurosci 7:279–308

Björklund A, Dunnett SB, Stenevi U, Lewis MR, Iversen SD (1980a) Reinnervation of the denervated striatum by substantia nigra transplants: functional consequences as revealed by pharmacological and sensorimotor testing. Brain Res 199:307–333

Björklund A, Schmidt RH, Stenevi U (1980b) Functional reinnervation of the neostriatum in the adult rat by use of intraparenchymal grafting of dissociated cell suspensions from the substantia nigra. Cell Tissue Res 212:39–45

Björklund A, Stenevi U, Dunnett SB, Iversen SD (1981) Functional reactivation of the de-afferented neostriatum by nigral transplants. Nature 289:497–499

Björklund A, Stenevi U, Schmidt RH, Dunnett SB, Gage FH (1983a) Intracerebral grafting of neuronal cell suspensions. I. Introduction and general methods of preparation. Acta Physiol Scand [Suppl] 522:1–7

Björklund A, Stenevi U, Schmidt RH, Dunnett SB, Gage FH (1983b) Intracerebral grafting of neuronal cell suspensions. II. Survival and growth of nigral cell suspensions implanted in different brain sites. Acta Physiol Scand [Suppl] 522:9–18

Björklund A, Lindvall O, Isacson O, Brunding P, Wictorin K, Strecker RE, Clarke DJ, Dunnett SB (1987) Mechanisms of action of intracerebral neural implants: studies on nigral and striatal grafts to the lesioned striatum. Trends Neurosci 10:509

Bohn MC, Cupit L, Marciano F, Gash DM (1987) Adrenal medulla grafts enhance recovery of striatal dopaminergic fibers. Science 237:913–916

Bolam JP, Freund TF, Björklund A, Dunnett SD, Smith AD (1987) Synaptic input and local output of dopaminergic neurons in grafts that functionally reinnervate the host neostriatum. Exp Brain Res 68:131–146

Cajal S Ramon Y (1928) Degeneration and regeneration in the nervous system. Oxford University Press, London

Chiodo LA, Antelman SM, Caggiula AR, Lineberry CG (1980) Sensory stimuli alter the discharge rate of dopamine (DA) neurons: evidence for two functional types of DA cells in the substantia nigra. Brain Res 189:544–549

Clarke DJ, Dunnett SB, Isacson O, Sirinathsinghji DJS, Björklund A (1988) Striatal grafts in rats with unilateral neostriatal lesions. I. Ultrastructural evidence of afferent synaptic inputs from the host nigrostriatal pathway. Neuroscience 24:791–801

Coyle JT, Schwarcz R (1976) Lesion of striatal neurones with kainic acid provides a model for Huntington's chorea. Nature 263:244–246

Deckel AW, Robinson RG, Coyle JT, Sanberg PR (1983) Reversal of long-term locomotor abnormalities in the kainic acid model of Huntington's disease by day 18 fetal striatal implants. Eur J Pharmacol 93:287–288

Deckel AW, Moran TH, Robertson RG (1986) Behavioral recovery following kainic acid lesions and fetal implants of the striatum occurs independent of dopaminergic mechanisms. Brain Res 363:383–385

Divac L, Markowitsch HJ, Pritzel M (1978) Behavioural and anatomical consequences of small intrastriatal injections of kainic acid in the rat. Brain Res 151:523–532

Dunnett SB, Iversen SD (1981) Learning impairments following selective kainic acid-induced lesions within the neostriatum of rats. Behav Brain Res 2:189–209

Dunnett SB, Iversen SD (1982) Neurotoxic lesions of ventrolateral but not anteromedial neostriatum in rats impair differential reinforcement of low rates (DRL) performance. Behav Brain Res 6:213–226

Dunnett SB, Björklund A, Stenevi U, Iversen SD (1981a) Behavioural recovery following transplantation of substantia nigra in rats subjected to 6-OHDA lesions of the nigrostriatal pathway. I. Unilateral lesions. Brain Res 215:147–161

Dunnett SB, Björklund A, Stenevi U, Iversen SD (1981b) Grafts of embryonic substantia nigra reinnervating the ventrolateral striatum ameliorate sensorimotor impairments and akinesia in rats with 6-OHDA lesions of the nigrostriatal pathway. Brain Res 229:209–217

Dunnett SB, Björklund A, Stenevi U, Iversen SD (1981c) Behavioural recovery following transplantation of substantia nigra in rats subjected to 6-OHDA lesions of the nigrostriatal pathway. II. Bilateral lesions. Brain Res 229:457–470

Dunnett SB, Björklund A, Schmidt RH, Stenevi U, Iversen SD (1983a) Intracerebral grafting of neuronal cell suspensions. IV. Behavioural recovery in rats with unilateral implants of nigral cell suspensions in different forebrain sites. Acta Physiol Scand [Suppl] 522:39–47

Dunnett SB, Björklund A, Schmidt RH, Stenevi U, Iversen SD (1983b) Intracerebral grafting of neuronal cell suspensions. V. Behavioural recovery in rats with bilateral 6-OHDA lesions following implantation of nigral cell suspensions. Acta Physiol Scand [Suppl] 522:39–47

Dunnett SB, Björklund A, Stenevi U (1983c) Dopamine-rich transplants in experimental parkinsonism. Trends Neurosci 6:266–270

Dunnett SB, Björklund A, Stenevi U (1983d) Transplant-induced recovery from brain lesions: a review of the nigrostriatal model. In: Wallace RD, Das GD (eds) Neural tissue transplantation research. Springer, Berlin Heidelberg New York, pp 191–216

Dunnett SB, Björklund A, Gage FH, Stenevi U (1985) Transplantation of mesencephalic dopamine neurons to the striatum of adult rats: In: Björklund A, Stenevi U (eds) Neural grafting in the mammalian CNS. Elsevier, Amsterdam, pp 451–469

Dunnett SB, Björklund A, Brundin P, Isacson O, Gage FH (1986a) Transplantation of dopamine cell suspensions to the dopamine-depleted neostriatum. In: Stern G (ed) Parkinson's disease. Chapman and Hall, London

Dunnett SB, Wishaw IQ, Jones GH, Isacson O (1986b) Effects of dopamine-rich grafts on conditioned rotation in rats with unilateral 6-OHDA lesions. Neurosci Lett 68:127–133

Dunnett SB, Whishaw IQ, Rogers DC, Jones GH (1987) Dopamine-rich grafts ameliorate whole body motor asymmetry and sensory neglect but not independent limb use in rats with 6-hydroxydopamine lesions. Brain Res 415:63–78

Dunnett SB, Hernandez TD, Summerfield A, Jones GH, Arbuthnott G (1988a) Graft-derived recovery from 6-OHDA lesions: specificity of ventral mesencephalic graft tissue. Exp Brain Res 71:411–424

Dunnett SB, Isacson O, Sirinathsinghji DHS, Clarke DJ, Björklund A (1988b) Striatal grafts in rats with unilateral neostriatal lesions. III. Recovery from dopamine dependent motor asymmetry and deficits in skilled paw reaching. Neuroscience 24:813–820

Fischer SJ, Young PM, Groves PM, Gage, FH (1988) Extracellular characterization of HRP-labelled neurons in dopamine-rich suspension grafts to the rat neostriatum. Soc Neurosci Abstr 14:292–297

Fray PJ, Dunnett SB, Iversen SD, Björklund A, Stenevi U (1983) Nigral transplants reinnervating the dopamine-depleted neostriatum can sustain intracerebral self-stimulation. Science 219:416–419

Freed WJ (1983) Functional brain tissue transplantation: reversal of lesion-induced rotation by intraventricular substantia nigra and adrenal medulla grafts, with a note on intracranial retinal grafts. Biol Psychiatry 18:1205−1267

Freed WJ, Perlow MJ, Karoum F, Seiger A, Olson L, Hoffer BJ, Wyatt RJ (1980) Restoration of dopaminergic function by grafting of fetal rat substantia nigra to the caudate nucleus: long-term behavioral, biochemical, and histochemical studies. Ann Neurol 8:510−519

Freed WJ, Morihisa JM, Spoor E, Hoffer BJ, Olson L, Seiger A, Wyatt RJ (1981) Transplanted adrenal chromaffin cells in rat brain reduce lesion-induced rotational behaviour. Nature 292:351−352

Freed WJ, Karoum F, Spoor E, Morihisa JM, Olson L, Wyatt RJ (1983) Catecholamine content of intracerebral adrenal medulla grafts. Brain Res 269:184−189

Freund T, Bolam JP, Björklund A, Stenevi U, Dunnett SB, Powell JF, Smith AD (1985) Efferent synaptic connections of grafted dopaminergic neurons reinnervating the host neostriatum: tyrosine hydroxylase immunocytochemical study. J Neurosci 5:603−616

Gage FH, Björklund A (1986) Neural grafting in the aged rat brain. Annu Rev Physiol 48:447−459

Gage FH, Stenevi U, Carlstedt T, Foster G, Björklund A, Aguayo AJ (1985) Anatomical and functional consequences of grafting mesencephalic neurons into a peripheral nerve "bridge" connected to the denervated striatum. Exp Brain Res 60:584−589

Gash DM, Collier TJ, Sladek JR Jr (1985) Neural transplantation: A review of recent developments and potential applications to the aged brain. Neurobiol Aging 6:131−150

Graybiel AM (1984) Neurochemically specified subsystems in the basal ganglia. Ciba Found Symp 107:114−144

Herman JP, Choulli K, Geffard M, Nadaud D, Taghzouti K, Le Moal M (1986) Reinnervation of the nucleus accumbens and frontal cortex of the rat by dopaminergic grafts and effects on hoarding behavior. Brain Res 372:210−216

Isacson O, Brundin P, Kelly PAT, Gage FH, Björklund A (1984) Functional neuronal replacement by grafted striatal neurones in the ibotenic acid-lesioned rat striatum. Nature 311:458−460

Isacson O, Brundin P, Gage FH, Björklund A (1985) Neural grafting in a rat model of Huntington's disease: progressive neurochemical changes after neostriatal ibotenate lesions and striatal tissue grafting. Neuroscience 16:799−817

Isacson O, Dunnett SB, Björklund A (1986) Graft-induced behavioural recovery in an animal model of Huntington's chorea. Proc Natl Acad Sci USA 83:2728−2732

Katzman R, Björklund A, Owman C, Stenevi U, West K (1971) Evidence for regeneration axon sprouting of central catecholamine neurons in the rat mesencephalon following electrolytic lesions. Brain Res 25:579−596

Kesslak J, Nieto-Sampedro M, Globus J, Cotman CW (1986) Transplants of purified astrocytes promote behavioral recovery after frontal cortex ablation. Exp Neurol 92:377−390

Ljungberg T, Ungerstedt U (1976) Reinstatement of eating by dopamine agonists in aphagic dopamine denervated rats. Physiol Behav 16:277−283

Lynch GS, Matthews DA, Mosko S, Parks T, Cotman CW (1972) Induced acetylcholinesterase-rich layer in dentate gyrus following entorhinal lesions. Brain Res 42:311−318

Mahalik TJ, Finger TE, Strömberg I, Olson L (1985) Substantia nigra transplants into denervated striatum of the rat: ultrastructure of graft and host interconnection. J Comp Neurol 240:60−70

Manthorpe M, Nieto-Sampedro M, Skaper SD, Lewis ER, Barin G, Longo FM, Cotman CW, Varon S (1983) Neuronotrophic activity in brain wounds of the developing rat. Correlation with implant survival in the wound cavity. Brain Res 267:47−56

Marshall JF, Gotthelf T (1979) Sensory inattention in rats with 6-hydroxydopamine-induced lesions of ascending dopaminergic neurons: apomorphine induced reversal of deficits. Exp Neurol 65:389−411

Marshall JF, Teitelbaum P (1977) New considerations in the neuropsychology of motivated behavior. In: Iversen LL, Iversen SD, Snyder SH (eds) Handbook of psychopharmacology, vol 7. Plenum, New York, pp 201−229

Marshall JF, Ungerstedt U (1976) Apomorphine-induced restoration of drinking to thirst challengers in 6-hydroxydopamine-treated rats. Physiol Behav 17:817−822

Marshall JF, Richardson JS, Teitelbaum P (1974) Nigrostriatal bundle damage and the lateral hypothalamic syndrome. J Comp Physiol Psychol 87:808–880

Mason ST, Fibiger HC (1978) Kainic acid lesions of the striatum: behavioural sequelae similar to Huntington's chorea. Brain Res 155:313–329

Nauta WJH, Domesick VB (1984) Afferent and efferent relationships of the basal ganglia. Ciba Found Symp 107:3–23

Nieoullon A, Cheramy A, Glowinski J (1977) Nigral and striatal dopamine release under sensory stimuli. Nature 269:349–351

Nieto-Sampedro M, Lewis ER, Cotman CW, Manthorpe M, Skaper SD, Barbin G, Longo FM, Varon S (1982) Brain injury causes a time-dependent increase in neuronotrophic activity at the lesion site. Science 217:860–861

Nieto-Sampedro M, Manthorpe M, Barbin G, Varon S, Cotman CW (1983) Injury-induced neuronotrophic activity in adult rat brain: correlation with survival of delayed implants in the wound cavity. J Neurosci 3:2219–2229

Perlow MJ, Freed WJ, Hoffer BJ, Seiger A, Olson L, Wyatt RJ (1979) Brain grafts reduce motor abnormalities produced by destruction of nigrostriatal dopamine system. Science 204:643–647

Pritzel M, Isacson O, Brundin B, Wiklund L, Björklund A (1986) Afferent and efferent connections of striatal grafts implanted into the ibotenic acid lesioned rats. Exp Brain Res 65:112–126

Raisman G (1969) Neuronal plasticity in the septal nuclei of the adult brain. Brain Res 14:25–48

Rutherford A, Garcia-Munoz M, Dunnett SB and Arbuthnott GW (1987) Electrophysiological demonstration of host cortical inputs into striatal grafts. Neurosci Lett 83:275–281

Sanberg PR, Calderon SF, Garver DL, Norman AB (1987) Brain tissue transplants in an animal model of Huntington's disease. Psychopharmacol Bull 23:476–482

Sanberg PR and Henault MA (1986) Fetal striatal transplants restricted to lateral ventricles in striatal lesioned rats do not produce recovery of abnormal locomotion. Soc Neurosci Abstr 12:563

Schmidt RH, Björklund A, Stenevi U (1981) Intracerebral grafting of dissociated CNS tissue suspensions: a new approach for neuronal transplantation to deep brain sites. Brain Res 218:347–356

Schmidt RH, Ingvar M, Lindvall O, Stenevi U, Björklund A (1982) Functional activity of substantia nigra grafts reinnervating the striatum: neurotransmitter metabolism and (^{14}C)-2-deoxy-D-glucose autoradiography. J Neurochem 38:737–748

Schmidt RH, Björklund A, Stenevi U, Dunnett SB, Gage FH (1983) Intracerebral grafting of neuronal cell suspensions. III. Activity of intrastriatal nigral suspension implants as assessed by measurements of dopamine synthesis and metabolism. Acta Physiol Scand [Suppl] 522:19–28

Schultz W (1982) Depletion of dopamine in striatum as an experimental model of Parkinsonism: direct effects and adaptive mechanism. Prog Neurobiol 18:121–166

Schwarcz R, Coyle JT (1977) Striatal lesions with kainic acid: neurochemical characteristics. Brain Res 127:235–249

Schwarz SS, Freed WJ (1986) Brain tissue transplantation in neonatal rats prevents a lesion-induced syndrome of adipsia, aphagia and akinesia. Exp Brain Res 65:449–454

Sirinathsinghji DJS, Dunnett SB, Isacson O, Clarke DJ, Kendrick K, Björklund A (1988) Striatal grafts in rats with unilateral neostriatal lesions. II. In vivo monitoring of GABA release in globus pallidus and substantia nigra. Neuroscience 24:803–811

Stenevi U, Björklund A, Dunnett SB (1980) Functional reinnervation of the denervated neostriatum by nigral transplants. Peptides [Suppl 1] 1:111–116

Strecker RE, Sharp T, Brundin P, Zetterström T, Ungerstedt U, Björklund A (1987) Autoregulation of dopamine release metabolism by intrastriatal nigral grafts as revealed by intracerebral dialysis. Neuroscience 22:169–178

Stricker EM, Zigmond MJ (1976) Recovery of function following damage to central catecholamine-containing neurons: a neurochemical model for the lateral hypothalamic syndrome. In: Sprague JM, Epstein AN (eds) Progress in physiological psychology. American Press, New York, pp 121–189

Strömberg I, Herrera-Marschitz M, Ungerstedt U, Ebendal T, Olson L (1985) Chronic implants of chromaffin tissue into the dopamine-denervated striatum. Effects of NGF on graft survival, fiber growth and rotational behavior. Exp Brain Res 60:335–349

Svendgaard N-AA, Björklund A, Stenevi U (1975) Regenerative properties of central monoamine neurons as revealed in studies using iris transplants as targets. Adv Anat Embryol Cell Biol 51:1–77

Ungerstedt U (1971a) Striatal dopamine release after amphetamine or nerve degeneration revealed by rotational behavior. Acta Physiol Scand [Suppl] 367:49–68

Ungerstedt U (1971b) Post-synaptic supersensitivity after 6-hydroxydopamine induced degeneration of the nigro-striatal dopamine system. Acta Physiol Scand [Suppl] 367:69–93

Ungerstedt U (1971c) Adipsia and aphagia after 6-hydroxydopamine induced degeneration of the nigro-striatal dopamine system. Acta Physiol Scand [Suppl] 367:95–122

Wictorin K, Björklund A (1989) Connectivity of striatal grafts implanted into the ibotenic acid lesioned striatum. II. Cortical afferents. Neuroscience (in press)

Wictorin K, Isacson O, Fischer W, Nothias F, Peschanski M, Björklund A (1988) Connectivity of striatal grafts implanted into the ibotenic acid-lesioned striatum. I. Subcortical afferents. Neuroscience 27:547–562

Wictorin K, Simerly RB, Isacson O, Swanson LW, Björklund A (1989a) Connectivity of striatal grafts implanted into the ibotenic acid lesioned striatum. III. Efferent projecting graft neurons and their relation to host afferents within the grafts. Neuroscience (in press)

Wictorin K, Clarke DJ, Bolam JP, Björklund A (1989b) Host corticostriatal fibres establish synaptic connections with grafted striatal neurons in the ibotenic acid lesioned striatum. Eur J Neurosci (in press)

Wilson CJ, Emson P, Feler C (1987) Electrophysiological evidence for the formation of a corticostriatal pathway in neostriatal tissue grafts. Soc Neurosci Abstr 13:11

Winn P, Tarbuck A, Dunnett SB (1984) Ibotenic acid lesions of the lateral hypothalamus: comparison with the electrolytic lesion syndrome. Neuroscience 12:225–240

Zetterström T, Brundin P, Gage FH (1986) Spontaneous release of dopamine from intrastriatal nigral grafts as monitored by the intracerebral dialysis technique. Brain Res 362:344–349

Zigmond MJ, Stricker EM (1972) Deficits in feeding behavior after intraventricular injection of 6-hydroxydopamine in rats. Science 177:1211–1213

Intracerebral Grafts of Dopaminergic Neurons: A Discussion of Their Functional Effects and Mechanisms of Action

J. P. Herman, K. Choulli, N. Abrous, and *M. Le Moal*

Summary

Studies performed in the past few years indicate that the intracerebral implantation of dopaminergic neurons can lead to a partial or total disappearance of a number of behavioral deficits elicited by the prior lesion of the dopaminergic system of the host. However, the functional action of these grafts also presents some limitations and unwanted characteristics: absence of recovery in some behavioral tests, exaggerated pharmacological responses, etc. It is suggested that the implanted neurons do not exert their functional effects solely through the restitution of normal physiological mechanisms, but rather by a mixture of normal and nonphysiological mechanisms. The detailed analysis of the pattern of recovery mediated by the implanted neurons indicates that, among the possible nonphysiological mechanisms, the lack of interneuronal regulations of the grafted neurons and the existence of abnormal postsynaptic mechanisms might indeed play a role in these limitations. The understanding of these mechanisms will be crucial when considering the application of intracerebral grafting to human therapy.

Introduction

The use of intracerebral neuronal grafting has been frequently discussed in the past few years as a potential therapeutic approach to human neurodegenerative diseases, and especially to Parkinson's disease and Alzheimer's disease (see also Björklund et al., this volume). As far as the latter examples are concerned, the neuronal populations considered for implantation (dopaminergic and cholinergic neurons, respectively) have several common characteristics: largely divergent and diffuse projections innervating very large telencephalic areas, a functional role which seems to be one of general modulation rather than precise computation, and a rather slow post-synaptic action. On the basis of these similarities these systems have been grouped together – with some other neuronal systems – among the so-called global systems (Sotelo and Alvarado-Mallart 1987) or class II systems (Strange 1988).

While there are relatively few reports of results concerning the functional effects of intracerebral implants of cholinergic neurons, recently dopaminergic grafts have been extensively studied. On the basis of the similarities mentioned above it can be assumed, however, that some of the conclusions reached through the study of implants of dopaminergic neurons – which are among the most extensively studied grafts – can be extended to grafts of other kinds of neurons belonging to the class of global systems and among these, to cholinergic neurons.

The ability of intracerebral implants of dopaminergic neurons to compensate for postlesional functional deficit in animals was first reported 10 years ago (Björklund and Stenevi 1979; Perlow et al. 1979). These studies marked the starting point of a series of investigations exploring the range of postlesion behavioral deficits which could be influenced by such implants (Björklund et al. 1980a, b; Freed et al. 1980; Dunnett et al. 1981a, b, c, 1983a, b, 1984, 1986, 1987; Freed 1983; Nadaud et al. 1984; Herman et al. 1985, 1986, 1988a; Brundin et al. 1986; Choulli et al. 1987a, b; Schwartz and Freed 1987).

Table 1. Influence of intracerebral grafts of dopaminergic neurons on postlesion behavioral deficits

	Recovery (total or partial)	Recovery with stimulation	Detrimental effects
A9 Lesion			
Receptor hypersensitivity	+ [2,7,10,14,15,16,17,21]		
Amphetamine-induced rotation	+ [1,2,3,4,7,10,14,15,16,17]		+ [15,17]
Sensorimotor orientation	+ [8,9,10,14]		
Sensorimotor limb use	+ [10,14]		
T-maze side bias	+ [7,8,10]		
Conditioned rotation	+ [13]		+ [13]
Paw preference	− [14]		
Akinesia, adult graft	− [8,9,11]		
Akinesia, neonatal graft	+ [22]		
Aphagia, adult graft	− [2,8,9,11]		
Aphagia, neonatal graft	+ [22]		
Adipsia, adult graft	− [2,8,9,11]		
Adipsia, neonatal graft	+ [22]		
A10 Lesion-Terminals			
Receptor hypersensitivity	+ [6,17,19]		
Amphetamine-induced locomotion	+ [17,18,19]		+ [17,19]
Exploration		+ [19]	+ [6]
Hoarding		+ [19]	+ [6]
Spatial orientation	− [5]		
SIP	− [5]		
A10 Lesion-cell bodies			
Receptor hypersensitivity	+ [12,19]		
Amphetamine-induced locomotion	+ [5,12,19,20]		+ [12,19]
Spontaneous locomotor activity			+ [12,19]
Exploration	− [19]		
Hoarding	− [19]		
Spatial orientation	− [5]		
SIP	− [5]		

1: Björklund and Stenev 1979; 2: Björklund et al. 1980a; 3: Björklund et al. 1980b; 4: Brundin et al. 1965; 5: Choulli et al. 1987a; 6: Choulli et al. 1987b; 7: Dunnett et al. 1981a; 8: Dunnett et al. 1981b; 9: Dunnett et al. 1981c; 10: Dunnett et al. 1983a; 11: Dunnett et al. 1983b; 12: Dunnett et al. 1984; 13: Dunnett et al. 1986; 14: Dunnett et al. 1987; 15: Freed 1983; 16: Freed et al. 1980; 17: Herman et al. 1985; 18: Herman et al. 1986; 19: Herman et al. 1988a; 20: Nadaud et al. 1984; 21: Perlow et al. 1979; 22: Schwartz and Freed 1987

The main results of these studies are summarized in Table 1. It appears that the intracerebral implantation of embryonic dopaminergic neurons leads indeed to a partial or total disappearance of a number of behavioral deficits elicited by the prior lesion of the dopaminergic systems of the host. However, a number of limitations and unwanted aspects of these functional effects have also been disclosed, such as the existence of exaggerated pharmacological responses (Herman et al. 1985), the lack of recovery for some deficits (e.g., aphagia and adipsia; Dunnett et al. 1981b, c, 1983b), the requirement for an exogenous stimulation of the grafted neurons to reveal a positive effect of the grafted neurons (e.g., in the case of deficit of hoarding behavior; Herman et al. 1986) and, in some models, the appearance of some new functional disturbances (e.g., conditioned rotation; Dunnett et al. 1986).

Several factors might explain, on a theoretical ground, these limitations. Some of these factors will be discussed below and evidence for their involvement in the limitations of the functional actions of implanted dopaminergic neurons will be presented.

Lack of Interneuronal Regulations

In experiments aimed at testing the behavioral effects of implants of dopaminergic neurons, these neurons are usually implanted into target areas of the normal dopaminergic system. The reason for this ectopic location is the limited growth capacity of neuronal processes within the adult CNS, which is manifested in the fact that dopaminergic neurons implanted in the mesencephalon, i.e., the site where the normal mesotelencephalic dopaminergic neurons are located, are not able to reinnervate distant target areas by sending axons across the gap constituted by nontarget regions lying between the mesencephalon and these targets (Björklund et al. 1983; Abrous et al. 1988; Herman et al. 1988c). The consequence of this limitation is that dopaminergic neurons must be implanted directly into the area to be reinnervated.

This nonphysiological location results in the reinnervation of the given target region by the implanted dopaminergic neurons (Björklund et al. 1983; Abrous et al. 1988). However, as a consequence of the limited growth mentioned above, it also renders unlikely the reinnervation of these neurons by the physiological afferents terminating in the mesencephalon and innervating the in situ mesencephalic dopaminergic neurons, even if these afferents are deprived of their normal targets by the prior lesion of the endogenous dopaminergic system of the host.

The lack of innervation should lead to the absence of modulation of the activity of the grafted dopaminergic neurons by physiological stimuli influencing the functioning of the normal, in situ mesotelencephalic dopaminergic neurons. In order to test this hypothesis, dopaminergic neurons were grafted into the nucleus accumbens of rats bearing a lesion of the endogenous dopaminergic neurons innervating this structure, and the activity of the grafted neurons was evaluated following the exposure of the animal to electric foot-shock stress (Herman et al. 1988c). In accordance with earlier results (Fadda et al. 1978; Herman et al. 1982) such a stress activated the dopaminergic terminals of the nucleus accumbens in control (non-

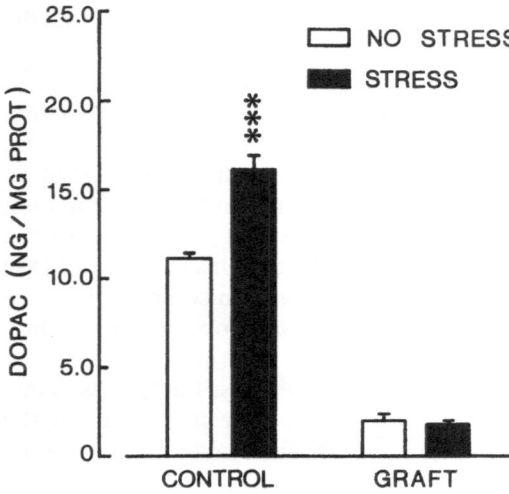

Fig. 1. Effect of exposure to electric foot-shock stress on DOPAC levels in the nucleus accumbens of control animals (*control* group) and animals with lesion of the mesocorticolimbic dopaminergic pathway and implantation of dopaminergic neurons into the nucleus accumbens (*graft* group). ***, $P < 0.005$, comparison with no stress conditions

lesioned) animals, as shown by the increase of dihydroxyphenylacetic acid (DOPAC) levels of this region (Fig. 1). However, no activation could be detected for the grafted neurons, thereby confirming the above hypothesis.

The importance of this lack of regulation in the action of the graft is suggested by the analysis of the functional recovery observed following the local lesion of the dopaminergic innervation of the nucleus accumbens (Choulli et al. 1987b). Such a lesion can be obtained by the intra-accumbens infusion of 6-hydroxydopamine, a procedure leading to the destruction of the dopaminergic terminals restricted to the nucleus accumbens, the anteromedial striatum, and the frontal cortex.

The evolution of some of the behavioral deficits provoked by this lesion was followed over 10 months and was compared with that observed in animals bearing an implant of dopaminergic neurons within the nucleus accumbens.

In the grafted animals the deficits persisted throughout the experiment (Fig. 2B), despite the presence in the nucleus accumbens of a rich dopaminergic reinnervation provided by the grafted neurons, approaching the innervation density observed in nonlesioned control animals. It must be noted that a transient recovery could be obtained, however, if the grafted dopaminergic neurons were stimulated by a low dose of amphetamine (Herman et al. 1986). On the contrary, the deficits were gradually attenuated in the lesioned animals and, by the end of the experimental period, were no longer apparent (Fig. 2A). However, the deficits could be reinstated at this time by performing a second lesion, similar to the first one (Choulli et al. 1987b). This latter result indicates that the recovery observed in the lesioned, nongrafted animals could be attributed to a dopaminergic reinnervation of the previously denervated area by mesencephalic neurons which had not undergone a retrograde degeneration following the first lesion. It must be stressed, however, that the histological analysis indicated that the density of this reinnervation was quite low.

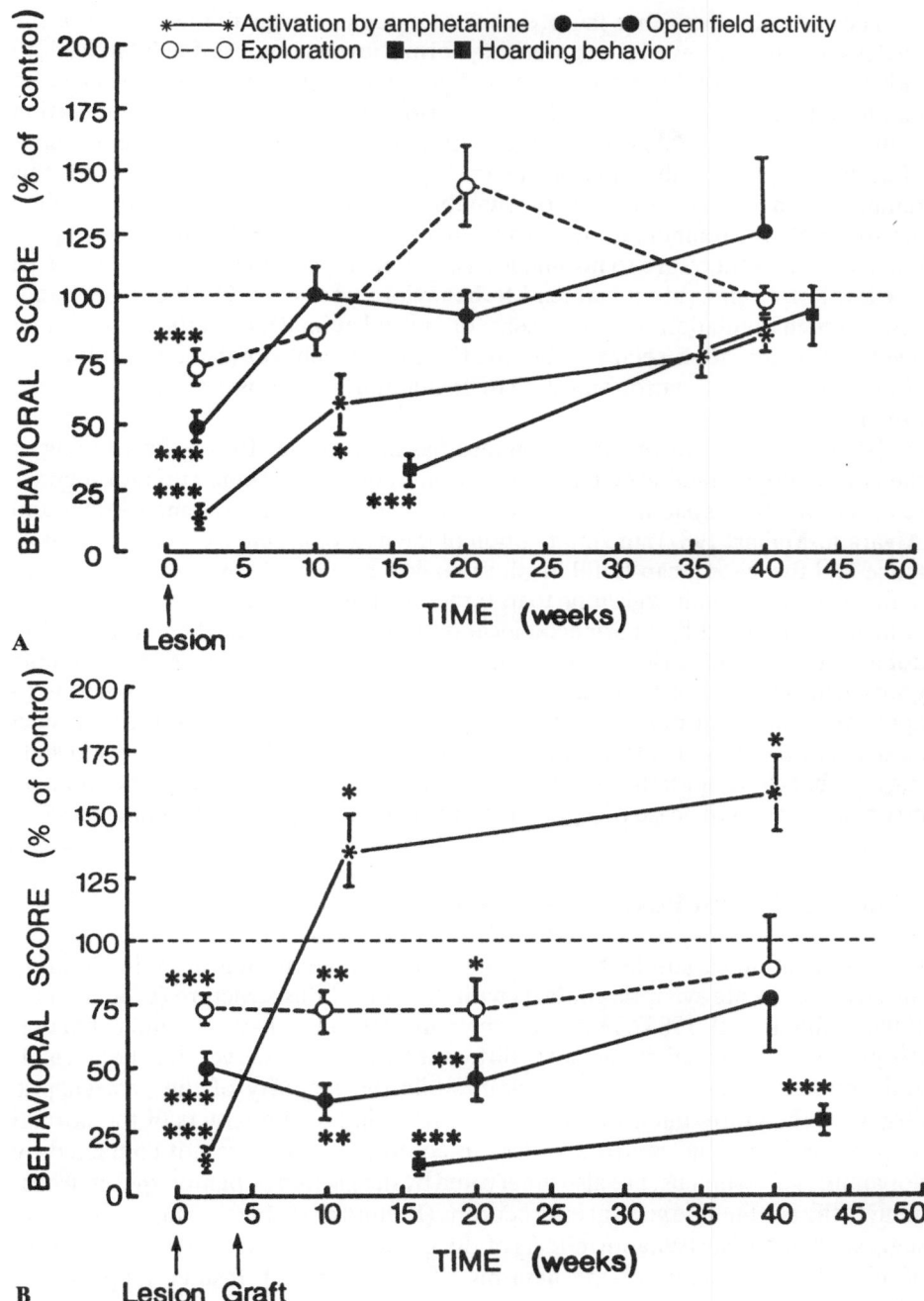

Fig. 2 A, B. Time-course of recovery of different behavioral deficits following the local lesion of the dopaminergic terminals of the nucleus accumbens (**A**) and a similar lesion followed by the implantation of dopaminergic neurons into the nucleus accumbens (**B**). Results are expressed as as percentage of the behavioral score of intact control animals. *, $P < 0.05$; **, $P < 0.01$; ***, $P < 0.005$, comparison with control animals. (From Choulli et al. 1987b)

These results show that the presence of a dopaminergic reinnervation of the nucleus accumbens, even if approaching normal levels, is not sufficient in itself to induce a recovery of these behaviors, and that some other conditions must also be met to obtain such a recovery. The comparison of the origin of the reinnervation in the two experimental groups suggests that this condition could be the presence of a physiological modulation of the activity of the neurons responsible for this reinnervation. This condition is presumably fulfilled in the case of the lesioned group, where the reinnervation is provided by neurons located in the mesencephalon, submitting therefore to normal interneuronal regulations. In this case even a low level of reinnervation can lead to behavioral recovery. On the other hand, with no such regulations being present for grafted animals, even the near-normal level of reinnervation seems to be insufficient for behavioral recovery and the effect of the grafted neurons must by further amplified to observe a positive functional effect.

Notice that the above conclusion also has implications for the functioning of the normal dopaminergic system. It has been frequently assumed that for this system, as for other systems belonging to the class of global systems (Sotelo and Alvarado-Mallart 1987), the modulation of the neuronal activity is of little importance and the system can fulfill its physiological role simply by providing a minimal, constant dopaminergic tone to its target structures. In fact it is this hypothesis (which is supported by pharmacological data, such as the beneficial action of L-dopa in Parkinson's disease) which lies behind the procedure of using ectopic grafts with the aim of inducing behavioral recovery. This hypothesis seems to apply to behavioral models in which spontaneous recovery mediated by such ectopic grafts exists (Table 1, first column). On the other hand, the above results suggest that for other behaviors the modulation of the neuronal activity could be essential, even in the case of a global system like the dopaminergic system.

Nonphysiological Postsynaptic Mechanisms

Previous anatomical studies have shown that dopaminergic neurons reinnervating the striatum create synaptic contacts with neurons of this structure (Freund et al. 1985; Mahalik et al. 1985). However, these investigations have also indicated that the pattern of connectivity is partly different from that observed for the normal, endogenous dopaminergic innervation: while the majority of the postsynaptic targets of the grafted neurons are of the same kind as the targets of the normal dopaminergic system, neurons which, in control animals, are not contacted by dopaminergic terminals, are also innervated by the grafted dopaminergic neurons, namely the cholinergic striatal interneurons (Freund et al. 1985). This observation suggests that the postsynaptic effects of the grafted neurons can involve a mixture of normal and abnormal mechanisms. Two groups of results support this hypothesis.

Reinstatement of Dopaminergic Control over Striatal Cholinergic Interneurons

It is known that the nigrostriatal dopaminergic pathway exerts an inhibitory control over the striatal cholinergic interneurons (for review see Lehman and Langer 1983). A recent study, using an in vitro superfusion of striatal slices, examined whether dopaminergic neurons have a similar action following grafting into the striatum deprived of its endogenous dopaminergic input (Herman et al. 1988b). On the basis of various pharmacological approaches it could be established that the grafted neurons are indeed able to reinstate a dopaminergic control over the striatal cholinergic neurons (Herman et al. 1988b). The existence of such control is shown, for example, by the decrease of acetylcholine release from striatal slices following the addition of amphetamine to the superfusion medium (Fig. 3). This decrease reflects the action of dopamine released by the drug from the dopaminergic terminals lying within the slice. While no such decrease existed for striatal slices obtained from lesioned animals, the acetylcholine outflow from slices containing the grafted dopaminergic neurons decreased following the addition of amphetamine to the superfusion medium, as it did in the case of slices from control striatum.

Strikingly, on all parameters examined the dopaminergic control exerted by the grafted neurons over the cholinergic neurons was of comparable magnitude to that observed for the normal nigrostriatal system, despite the fact that the overall density of striatal dopaminergic innervation, judged on the basis of the dopamine content of the striatum, was much lower (around 10% of the control values) in the case of the graft. This observation suggests that the functional effectiveness, with respect to the cholinergic neurons, of the terminals formed by the grafted neurons was higher than that of the normal nigrostriatal terminals. This higher effectiveness, compensating for the lower overall innervation density, could be related to the nonphysiological termination pattern mentioned above, i.e., the axosomatic terminals on the surface of the cholinergic neurons. These terminals, by their strategic location, could influence the cholinergic neuron more effectively than the presumably presynaptic terminals of the normal system (Lehman and Langer 1983; Freund et al. 1985).

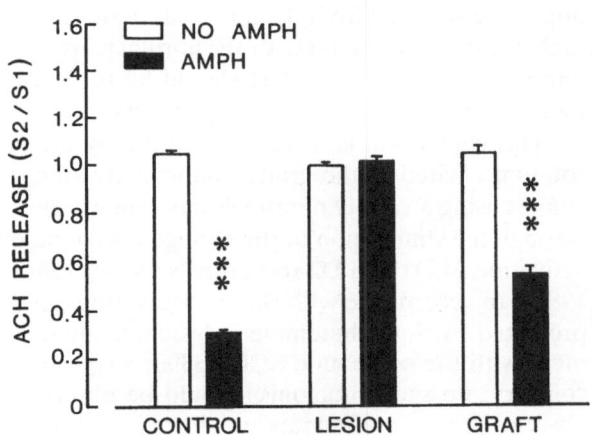

Fig. 3. Effect of *dl*-amphetaime (AMPH; 3 *M*) on in vitro 3H-acetylcholine *(ACH)* release from striatal slices. Slices were obtained from control striata *(control* group), striata with their nigrostriatal dopaminergic afferents lesioned *(lesion* group) and striata containing dopaminergic neurons implanted following the destruction of the nigrostriatal pathway *(graft* group). ***, $P < 0.005$, comparison with the release found in the absence of amphetamine. For details see Herman et al. 1988b

This example suggests that implanted dopaminergic neurons are able to exert a normal postsynaptic action within the target structure. Notice, however, that if the hypothesis concerning the involvement of the axosomatic terminals in this action is verified, it would mean that, even in the case of restoration by the graft of a globally normal postsynaptic regulation, the precise local mechanisms by which this regulation is brought about could be different from the mechanisms existing for the normal endogenous system.

Exaggerated Pharmacological Responses Following Intrastriatal Grafting

The analysis of the effect of intrastriatal grafts on rotational behavior following the unilateral lesion of the nigrostriatal system also suggests the existence of some abnormal postsynaptic mechanisms. Lesioned animals display, in response to amphetamine, an ipsilateral rotation, corresponding to the release of dopamine provoked by the drug in the nonlesioned striatum (Ungerstedt 1971), which disappears in animals bearing a graft of dopaminergic neurons in the denervated striatum (Björklund and Stenevi 1979; Björklund et al. 1980a, b; Freed et al. 1980; Dunnett et al. 1981a, 1983a, 1987; Freed 1983; Herman et al. 1985; Brundin et al. 1986). Nevertheless, the action of the graft cannot be equated to the reinstatement of a normal behavior. This is shown by the fact that the grafted animals display, following the injection of amphetamine, a marked *contra*lateral rotation which, moreover, lasts for 5 h (Fig. 4). This rotation is related to the release of dopamine from the grafted neurons, as shown by its abolition by a pretreatment of the animals with haloperidol, a dopamine receptor antagonist (Herman et al. 1985).

This paradoxical action was attributed to an enhanced sensitivity of the grafted neurons to amphetamine. It was hypothetized that, as a consequence of deficient interneuronal regulations (see above), the grafted neurons would release more dopamine under the action of amphetamine than the normal dopaminergic terminals in the nonlesioned striatum, and this higher dopaminergic tone in the grafted striatum would then lead to the contralateral rotation observed (Herman et al. 1985). However, in the meantime both in vivo (Zetterström et al. 1986) and in vitro (Herman et al. 1988b) measurements have shown that the amount of dopamine released from the terminals of the grafted neurons is lower, rather than higher, than that observed for the normal nigrostriatal terminals. This observation implies that the above effect should be related to some abnormal *post*synaptic mechanisms rather than *pre*synaptic ones.

This conclusion is further supported by the pharmacological analysis of the rotation elicited in the grafted animals by amphetamine. In the past few years studies using a variety of models have shown that, to elicit a behavioral response through the stimulation of the endogenous dopaminergic system, a simultaneous activation of D1 and D2 receptors is usually required (Waddington and O'Boyle 1988). In accordance with these observations, the ipsilateral rotational response provoked by d-amphetamine in lesioned animals could be blocked by pretreatment with the compound SCH 23390, a specific D1 antagonist (Fig. 5). On the contrary, no such antagonism could be observed for the contralateral rotation observed for the grafted rats and, in fact, a tendency for an increased response was

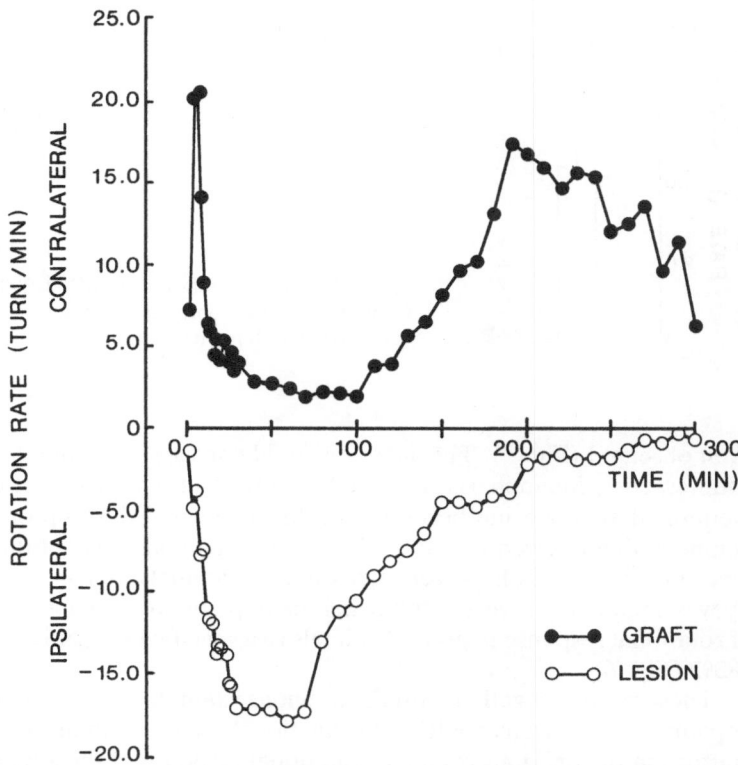

Fig. 4. Rotational response to *d*-amphetamine (5 mg/kg i.p. at *t* = 0 min) following unilateral lesion of the the nigrostriatal dopaminergic pathway (*lesion* group) or a similar lesion followed by the implantation of dopaminergic neurons into the denervated striatum (*graft* group)

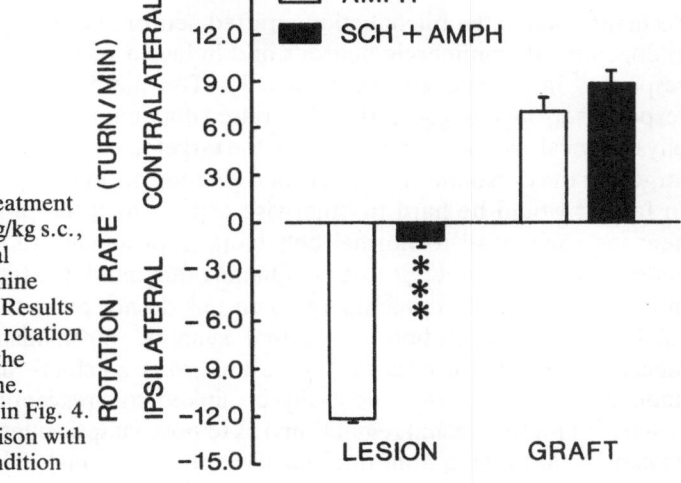

Fig. 5. Effect of a pretreatment with SCH 23389 (0.1 mg/kg s.c., 30 min) on the rotational response to d-amphetamine (AMPH; 5 mg/kg i.p.). Results correspond to the mean rotation rate over 1 h following the injection of amphetamine. Groups are the same as in Fig. 4. ***, $P < 0.005$, comparison with the no pretreatment condition

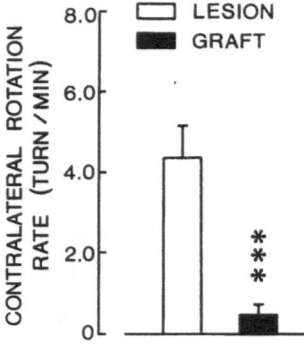

Fig. 6. Rotational response to SKF 38393 (2.5 mg/kg i.p.). Groups are the same as in Fig. 4. ***, $P < 0.005$, comparison with the lesion group

even observed (Fig. 5). This increase could correspond to the diminution of the counteracting dopaminergic tone in the contralateral, intact striatum. It should be mentioned that the inefficacy of the blockade of D1 receptors for inhibiting amphetamine-induced rotation in the grafted animals was not due to the persistence of D1-receptor hypersensitivity in these animals. In fact, this hypersensitivity was completely reversed following the implantation, as shown by the absence of rotational response in grafted animals to a specific D1 agonist, compound SKF 38393 (Fig. 6).

These results, together with the previous example, suggest that the implanted dopaminergic neurons could elicit functional effects – inhibition of cholinergic neurons in the previous example, rotational response of the animals in the latter – which involve postsynaptic mechanisms different from those through which the normal dopaminergic system induces a similar response. Future experiments should aim at elucidating more precisely these mechanisms.

Conclusions

As mentioned in the Introduction, grafted neurons can compensate for the loss of endogenous dopaminergic neurons and induce a recovery of *normal* behavioral responses in several models (Table 1). The mere existence of these normal responses strongly suggests that the grafted dopaminergic neurons restore normal, physiological synaptic mechanisms in the target tissue by innervating physiological targets of the dopaminergic system and eliciting normal responses in these targets. In fact, it would be hard to otherwise explain how the presence of the grafted neurons could lead to normal behaviors, reproducing those subserved by the endogenous dopaminergic system. On the other hand, the results presented above indicate that grafted dopaminergic neurons do not exert their functional action solely by the restitution of normal synaptic mechanisms; some abnormal mechanisms are also present. These abnormal mechanisms, although not well understood as yet, could as easily be linked to presynaptic factors (e.g., the absence of interneuronal regulations) as to postsynaptic ones (innervation of non-physiological targets, abnormal postsynaptic reactions). The existence of these abnormal mechanisms could explain some of the limitations of the functional effects of the grafted dopaminergic neurons. However, some other factors could also be involved.

First, it should be remembered that the current techniques of intracerebral grafting involve the implantation of a mixed population of cells in which the population of interest represents only a small percentage. Concerning grafts of embryonic mesencephali, such as were used in the above experiments, the proportion of dopaminergic neurons relative to the total intragraft neuronal population was estimated to be around 0.2% (Brundin and Björklund 1987). While the influence of these other neuronal populations is not known, their participation in some of the functional effects of the grafts cannot be excluded.

A second factor is related to the fact that the usual experimental paradigm for intracerebral grafting, at least in the case of adult hosts, involves the making of a lesion of the endogenous dopaminergic system and the implantation, some weeks later, of dopaminergic neurons into the area to be reinnervated. However, in this time interval the initial lesion could also provoke various secondary modifications in the host CNS (sprouting, transsynaptic degeneration, etc.) which might persist even after grafting and lead to modified behavioral responses. Notice that this aspect can be even more important in the case of slow neurodegenerative processes such as those for which the therapeutic use of intracerebral implantation is considered.

Finally, it should be mentioned that the persistence of postlesional deficits following grafting can also be due to the inadequate site of implantation (Dunnett et al. 1981c) or the lack of reinnervation of the totality of the areas denervated by the initial lesion (Herman et al. 1988a).

Better knowledge of the precise mechanisms through which implanted dopaminergic grafts exert their functional effects could help to understand the origin of the present limitations of these functional effects. Furthermore, as mentioned in the Introduction, it can also be relevant for the understanding of the mechanisms of action of grafts of other classes of neurons, such as the cholinergic neurons. Based on this knowledge steps can be envisaged to try to overcome the present limitations of the functional action of intracerebral grafts, which is a prerequisite to their therapeutic application.

References

Abrous N, Vigny A, Calas A, Le Moal M, Herman JP (1988) Development of intracerebral dopaminergic grafts: a combined immunohistochemical and autoradiographic study of its time course and environmental influences. J Comp Neurol 273:26–41

Björklund A, Stenevi U (1979) Reconstruction of the nigrostriatal dopamine pathway by intracerebral nigral transplants. Brain Res 177:555–560

Björklund A, Dunnett SB, Stenevi U, Lewis ME, Iversen SD (1980a) Reinnervation of the denervated striatum by substantia nigra transplants: functional consequences as revealed by pharmacological and sensorimotor testing. Brain Res 199:307–333

Björklund A, Schmidt RH, Stenevi U (1980b) Functional reinnervation of the neostriatum in the adult rat by use of intraparenchymal grafting of dissociated cell suspensions from the substantia nigra. Cell Tissue Res 212:39–45

Björklund A, Stenevi U, Schmidt RH, Dunnett SB, Gage FH (1983) Intracerebral grafting of neuronal cell suspensions. II. Survival and growth of nigral cell suspensions implanted in different brain sites. Acta Physiol Scand [Suppl] 522:9–18

Brundin P, Björklund A (1987) Survival, growth and function of dopaminergic neurons grafted to the brain. In: Seil FJ, Herbet E, Carlson BM (eds) Neural regeneration. Prog Brain Res 71:293–308

Brundin P, Isacson O, Gage FH, Prochiantz A, Björklund A (1986) The rotating 6-hydroxy-dopamine-lesioned mouse as a model for assessing functional effects of neuronal grafting. Brain Res 366:346–349

Choulli K, Herman JP, Abrous A, Le Moal M (1987a) Behavioral effects of intraaccumbens transplants in rats with lesions of the mesocorticolimbic dopamine system. Ann NY Acad Sci 495:497–509

Choulli K, Herman JP, Rivet M, Simon H, Le Moal M (1987b) Spontaneous and graft-induced behavioral recovery after 6-hydroxydopamine lesion of the nucleus accumbens in the rat. Brain Res 407:376–380

Dunnett SB, Björklund A, Stenevi U, Iversen SD (1981a) Behavioural recovery following transplantation of substantia nigra in rats subjected to 6-OHDA lesion of the nigrostriatal pathway. I. Unilateral lesion. Brain Res 215:147–161

Dunnett SB, Björklund A, Stenevi U, Iversen SD (1981b) Behavioural recovery following transplantation of substantia nigra in rats subjected to 6-OHDA lesion of the nigrostriatal pathway. II. Bilateral lesion. Brain Res 229:457–470

Dunnett SB, Björklund A, Stenevi U, Iversen SD (1981c) Grafts of substantia nigra reinnervating the ventrolateral striatum ameliorate sensorimotor impairments and akinesia in rats with 6-OHDA lesions of the nigrostriatal pathway. Brain Res 229:209–217

Dunnett SB, Björklund A, Schmidt RH, Stenevi U, Iversen SD (1983a) Intracerebral grafting of neuronal cell suspensions. IV. Behavioural recovery in rats with unilateral 6-OHDA lesions following implantation of nigral cell suspensions in different forebrain sites. Acta Physiol Scand [Suppl] 522:29–37

Dunnett SB, Björklund A, Schmidt RH, Stenevi U, Iversen SD (1983b) Intracerebral grafting of neuronal cell suspensions. V. Behavioural recovery in rats with bilateral 6-OHDA lesions following implantation of nigral cell suspensions. Acta Physiol Scand [Suppl] 522:39–47

Dunnett SB, Bunch ST, Gage FH, Björklund A (1984) Dopamine-rich transplants in rats with 6-OHDA lesions of the ventral tegmental area. I. Effects on spontaneous and drug induced locomotor activity. Behav Brain Res 13:71–82

Dunnett SB, Whishaw IQ, Jones GH, Isacson O (1986) Effects of dopamine-rich grafts on conditioned rotations in rats with unilateral 6-hydroxydopamine lesions. Neurosci Lett 68:127–133

Dunnett SB, Whishaw IQ, Rogers DC, Jones GH (1987) Dopamine-rich grafts ameliorate whole body motor asymmetry and sensory neglect but not independent limb use in rats with 6-hydroxydopamine lesions. Brain Res 415:63–78

Fadda F, Argiolas A, Melis MR, Tissari AH, Onali PL, Gessa GL (1978) Stress-induced increase in 3,4-dihydroxyphenylacetic acid (DOPAC) levels in the cerebral cortex and in n. accumbens: reversal by diazepam. Life Sci 23:2219–2224

Freed WJ (1983) Functional brain tissue transplantation: reversal of lesion-induced rotation by intraventricular substantia nigra and adrenal medulla grafts, with note on intracranial retinal grafts. Biol Psychiatry 18:1205–1267

Freed WJ, Perlow MJ, Karoum F, Seiger Å, Olson L, Hoffer BJ, Wyatt RJ (1980) Restoration of dopaminergic function by grafting of fetal rat substantia nigra to the caudate nucleus: long-term behavioral, biochemical, and histochemical studies. Ann Neurol 8:510–519

Freund TF, Bolam JP, Björklund A, Stenevi U, Dunnett SB, Powell JF, Smith AD (1985) Efferent synaptic connections of grafted dopaminergic neurons reinnervating the host neostriatum: a tyrosine hydroxylase immunocytochemical study. J Neurosci 5:603–616

Herman JP, Guillonneau D, Dantzer R, Scatton B, Semerdjian-Roquier L, Le Moal M (1982) Differential effects of inescapable footshocks and of stimuli previously paired with inescapable footshocks on dopamine turnover in cortical and limbic areas of the rat. Life Sci 30:2207–2214

Hermann JP, Choulli K, Le Moal M (1985) Hyper-reactivity to amphetamine in rats with dopaminergic grafts. Exp Brain Res 60:521–526

Herman JP, Choulli K, Geffard M, Nadaud D, Taghzouti K, Le Moal M (1986) Reinnervation of the nucleus accumbens and frontal cortex of the rat by dopaminergic grafts and effects on hoarding behavior. Brain Res 372:210–216

Herman JP, Choulli K, Abrous N, Dulluc J, Le Moal M (1988a) Effects of intraaccumbens grafts on behavioral deficits induced by 6-OHDA lesions of the nucleus accumbens or A10 dopaminergic neurons: a comparison. Behav Brain Res 29:73–83

Herman JP, Lupp A, Abrous N, Le Moal M, Hertting G, Jackisch R (1988b) Intrastriatal dopaminergic grafts restore inhibitory control over striatal cholinergic neurons. Exp Brain Res 73:236–248

Herman JP, Rivet JM, Abrous N, Le Moal M (1988c) Intracerebral dopaminergic transplants are not activated by electrical footshock stress activating in situ mesocorticolimbic neurons. Neurosci Lett 90:83–88

Lehman J, Langer SZ (1983) The striatal cholinergic interneuron: synaptic target of dopaminergic terminals? Neuroscience 4:1105–1120

Mahalik TJ, Finger TE, Strömberg I, Olson L (1985) Substantia nigra transplants into denervated striatum of the rat: ultrastructure of grafts and host interconnections. J Comp Neurol 240:60–70

Nadaud D, Herman JP, Simon H, Le Moal M (1984) Functional recovery following transplantation of ventral mesencephalic cells in rats subjected to 6-OHDA lesions of the mesolimbic dopaminergic neurons. Brain Res 304:137–141

Perlow MI, Freed WJ, Hoffer BJ, Seiger Å, Olson L, Wyatt RJ (1979) Brain grafts reduce motor abnormalities produced by destruction of nigrostriatal dopamine system. Science 204:643–647

Schwartz SS, Freed WJ (1987) Brain tissue transplantation in neonatal rats prevents a lesion-induced syndrome of adipsia, aphagia and akinesia. Exp Brain Res 65:449–454

Sotelo C, Alvarado-Mallart RM (1987) Reconstruction of the defective cerebellar circuitry in adult Purkinje cell degeneration mutant mice by Purkinje cell replacement through transplantation of solid embryonic implants. Neuroscience 20:1–22

Strange PG (1988) The structure and mechanism of neurotransmitter receptors. Implication for the structure and function of the central nervous system. Biochem J 249:309–311

Ungerstedt U (1971) Striatal dopamine release after amphetamine or nerve degeneration revealed by rotation behaviour. Acta Physiol Scand [Suppl] 367:49–68

Waddington JL, O'Boyle KM (1988) The D1 dopamine receptor and the search for its functional role: from neurochemistry to behaviour. Rev Neurosci 1:157–184

Zetterström T, Brundin P, Gage FH, Sharp T, Isacson O, Dunnett SB, Ungerstedt U, Björklund A (1986) In vivo measurement of spontaneous release and metabolims of dopamine from intrastriatal nigral grafts using intracerebral dialysis. Brain Res 362:344–349

Purkinje Cell Replacement by Intracerebellar Neuronal Implants in an Animal Model of Heredodegenerative Ataxia: An Overview

C. Sotelo

Summary

Functional recovery through neuronal transplantation requires a precise synaptic integration of replaced neurons. Working on the cerebellum of adult Purkinje cell degeneration *(pcd)* mice, we have shown that the missing Purkinje cells can be replaced by grafting cerebellar primordia taken from normal mouse embryos. This replacement results from a selective invasion of Purkinje cells which leave the graft and penetrate into the mutant molecular layer, and from the terminal sprouting of host cerebellar axons providing the grafted Purkinje cells with a specific synaptic investment. The in vitro electrophysiological study of these grafted neurons discloses:

1. typical all-or-none climbing fiber responses;
2. graded parallel fiber EPSPs; and
3. inhibitory postsynaptic potentials, all with characteristics similar to those recorded in control mice.

Furthermore, immunocytochemical studies with Purkinje cell markers have shown that, when grafted, Purkinje cells are located in molecular layer regions no further than 0.6 mm from the host deep nuclei, they are able to send appropriate projections, and to establish synaptic contacts on nuclear neurons, partially reconstructing the nucleocortical projection. These results point to the possibility of neuronal replacement by specific synaptic integration of the grafted neurons into the deficient cerebellar circuitry, conditions needed for functional restoration in systems connected in a point-to-point manner.

Introduction

In addition to the still remote and uncertain therapeutic use of neuronal grafting in brain repair, this experimental approach is actually an excellent tool with which to unravel some questions related to brain development and plasticity. Indeed, it offers a unique experimental situation, that of bringing together immature and mature neural components, allowing researchers to determine whether or not neuronal development can continue between chronologically different partners, and to analyze the nature of the cellular interactions that can take place between them. The expected result of these interactions would be the synaptic integration of the grafted neurons into neuronal networks of the host, leading to the restoration of the structural integrity of the deficient brain.

Over the years, the biological material we have used to determine the degree of restoration following neural grafting has been the cerebellum of mice, affected by a neurological mutation, that has Purkinje cells (the pivotal element of the cerebellar cortex) as primary cellular target: the Purkinje cell degeneration (*pcd* mouse; Mullen 1977). In the *pcd* mouse (Mullen et al. 1976) the cerebellum develops normally until the end of the 2nd postnatal week, when Purkinje cells begin to degenerate. By P45 virtually all of this category of neurons has disappeared (Wassef et al. 1987), and the histopathology of the cerebellum resembles that which characterizes some familial heredodegenerative ataxia in man (Holmes 1907).

In order to repair the cerebellar circuit of the *pcd* mouse, the Purkinje cell replacement needs to achieve three essential prerequisites:
1. From all the postmitotic neurons and progenitors present in the grafts, only Purkinje cells must leave the implanted cellular mass and move to the correct position previously occupied by the missing cells. Indeed, the restoration of the cortical circuits implies that the embryonic Purkinje cells are able to integrate themselves, as the missing link, into the host deficient network.
2. Once they reach proper locations, the grafted Purkinje cells must not only follow their differentiation program and build up their dendritic trees, but they must also provoke the sprouting of host axon terminals recapitulating their normal synaptogenesis with specific host afferents. These developmental events would lead to the synaptic integration of the grafted Purkinje cells into the deficient cortical circuitry.
3. Finally, no functional improvement can be accomplished if the disrupted corticonuclear projection is not reestablished. For that, the grafted neurons need to grow axons that, navigating throughout the adult host cerebellum, could reach their appropriate targets in the deep nuclei, and synaptically contact proper host nuclear neurons.

Material and Methods

In the experiments reported here, donor tissue was taken from isogeneic cerebellar primordia of 12-day-old (E12) mouse embryos. We used as hosts 3- to 4-month-old homozygous *pcd* mice. Two different grafting procedures were performed: either the cerebellar primordia were mechanically dissociated in tissue culture medium and the cell suspensions used for grafting, or the primordia were sliced into small (1.0 to 2.0 mm^3) pieces, and each individual piece was used in solid implantations. In both instances, the cell suspensions or the solid grafts were injected, using small glass pipettes, at variable depths within the parenchyma of the host cerebellum. The results obtained with the two procedures were almost identical (Sotelo and Alvarado-Mallart 1986, 1987a).

Results

By immunohistochemically staining Purkinje cells (using selective markers – either an antibody against calbindin or against cyclic GMP-dependent protein kinase), we have been able to identify grafted Purkinje cells in the host cerebellar 2 to 4 months after implantation (Figs. 1 and 2). The ultrastructural and electro-physiological analyses of the synaptology of these grafted Purkinje cells have allowed us to conclude that all three of the prerequisites discussed above can be fulfilled, as summarized below.

1. All the grafts succeed in providing Purkinje cells to the mutant cerebellum. The majority of these grafted cells move out and spread on both sides of the injection track to the nearby molecular layer parenchyma. In the largest transplants, the diameter of the spread is about 1.4 mm, suggesting that Purkinje cells migrate within an adult cerebellar parenchyma for distances of about 700 μm. The immunohistochemical material showed that these neurons are attracted only to the defective molecular layer, since they were never encountered outside this layer. The first prerequisite for neuronal replacement is fulfilled: the Purkinje cell dendritic trees occupy a proper location and acquire a roughly normal

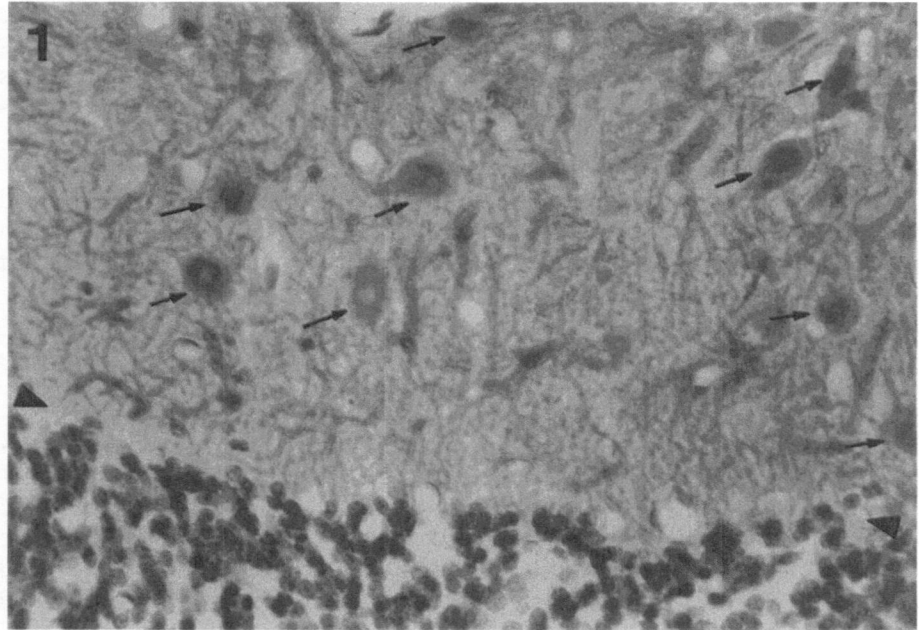

Fig. 1. Light micrograph of a grafted cerebellum 2 months after implantation. Grafted Purkinje cells *(arrows)* have been visualized with immunostaining using anticalbindin antibodies. Their cell bodies spread throughout the molecular layer, with the exception of its interface with the granular layer *(arrowhead)*, where they are normally situated. Note that the dendritic trees are composed of thick proximal branches and thin distal branches that occupy the whole depth of the molecular layer but that do not penetrate the granular layer; 10-μm thick paraffin, saggital section counterstained with cresyl violet. ×270

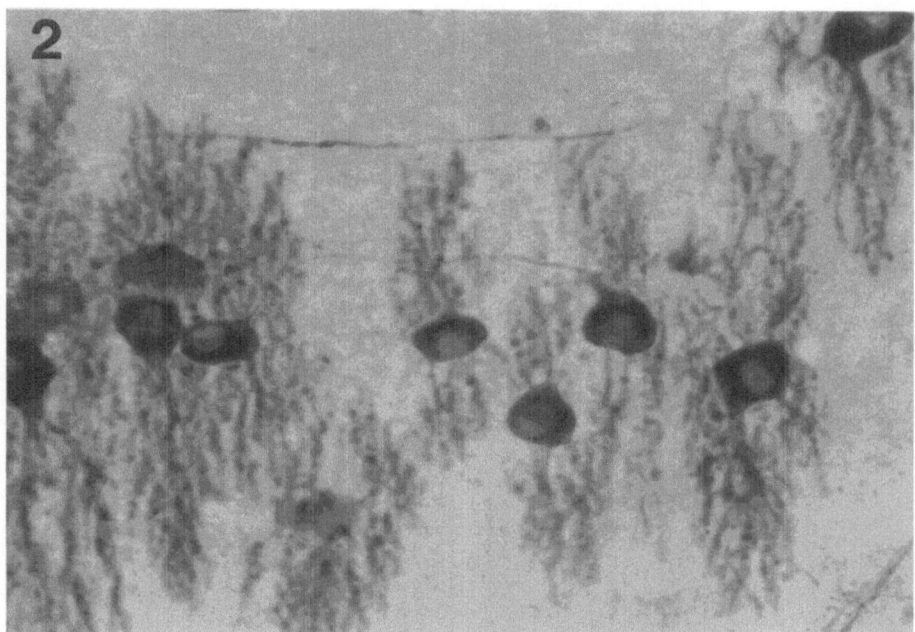

Fig. 2. Light micrograph of a grafted cerebellum 3 months after implantation, cut in the coronal plane. The cell bodies of the grafted Purkinje cells are distributed at different depths within the molecular layer. Note that the dendritic trees are flattened, oriented perpendicular to the bundles of host parallel fibers; 20-μm thick frozen section without counterstaining. Calbindin immunostaining. ×450

organization with proximal and distal compartments and monoplanar disposition (Figs. 1 and 2).

2. All grafted Purkinje cells invading the host molecular layer exhibit a similar synaptic investment. The ectopically located Purkinje cell somata receive inputs (Fig. 3) qualitatively similar to those of control Purkinje cells. However, one important difference exists: basket fiber-pinceau formations are absent around the initial segment of the Purkinje cell axon. Synaptic inputs to their dendrites also mimic normality (Figs. 4 and 5). Almost all primary thick dendritic branches receive climbing fiber varicosities (Fig. 4), which synapse on stubby spines. Their smooth dendritic segments receive axon terminals from molecular layer interneurons (Fig. 4), as in control cerebella. The distal dendritic compartment is composed of typical spiny branchlets (Fig. 5), studded with long-necked spines, which are contacted by axonal varicosities of the parallel fibers. From these data it can be inferred that the presence of embryonic, nonafferented Purkinje cells must provoke "terminal sprouting" of surviving target-deprived axon terminals in the host molecular layer.

The function of the neoformed synapses between the adult host neurons and the grafted Purkinje cells has been investigated using electrophysiologic

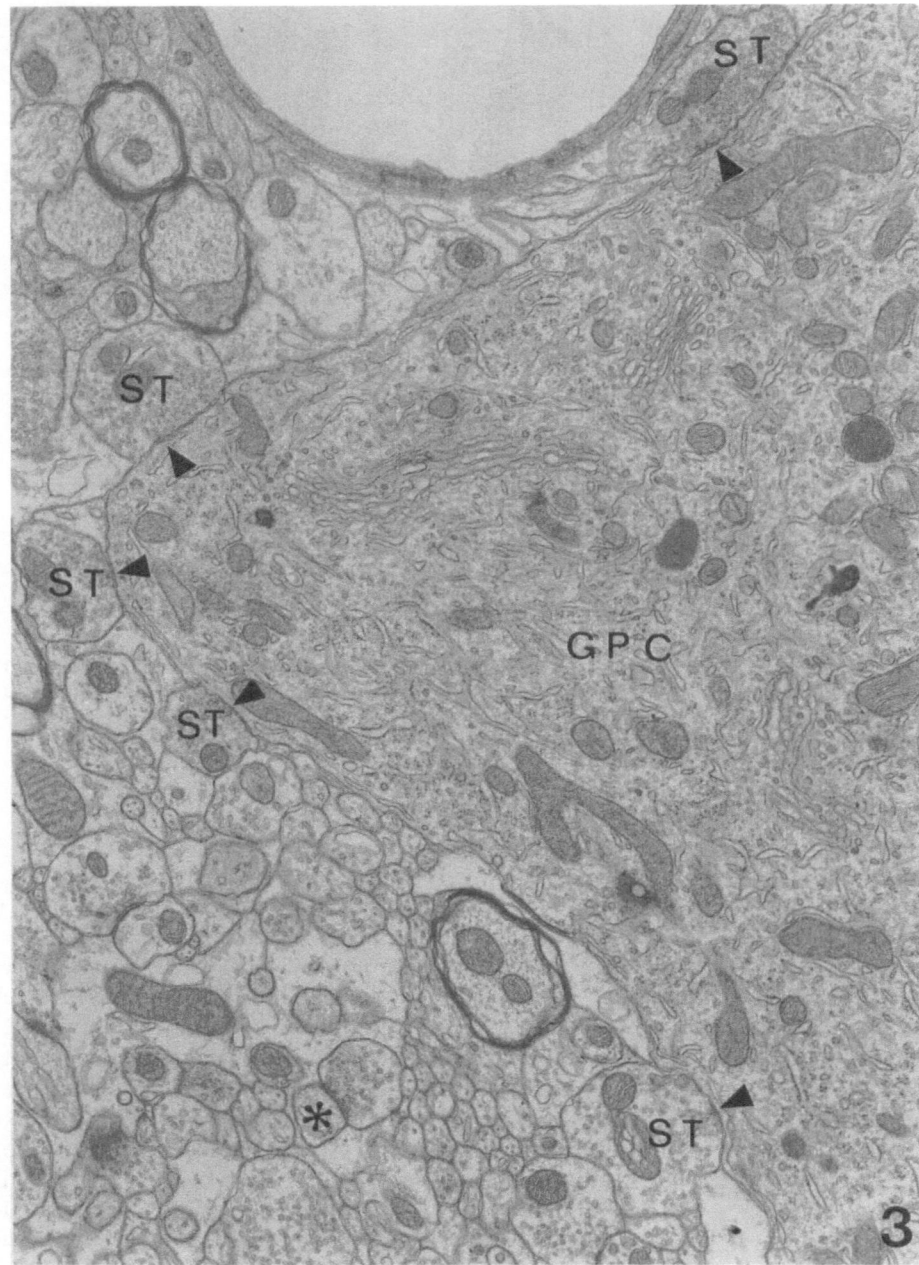

Fig. 3. Electron micrograph of a grafted Purkinje cell *(GPC)* 2 months after implantation. Five axon terminals belonging to stellate cell axons *(ST)* establish symmetrical synaptic contacts *(arrowheads)* with the somata of the grafted Purkinje cell. Note, in the neuropil, the presence of a parallel fiber varicosity synapsing on a distal spine *(asterisk)* of a Purkinje cell dendrite. ×17 000

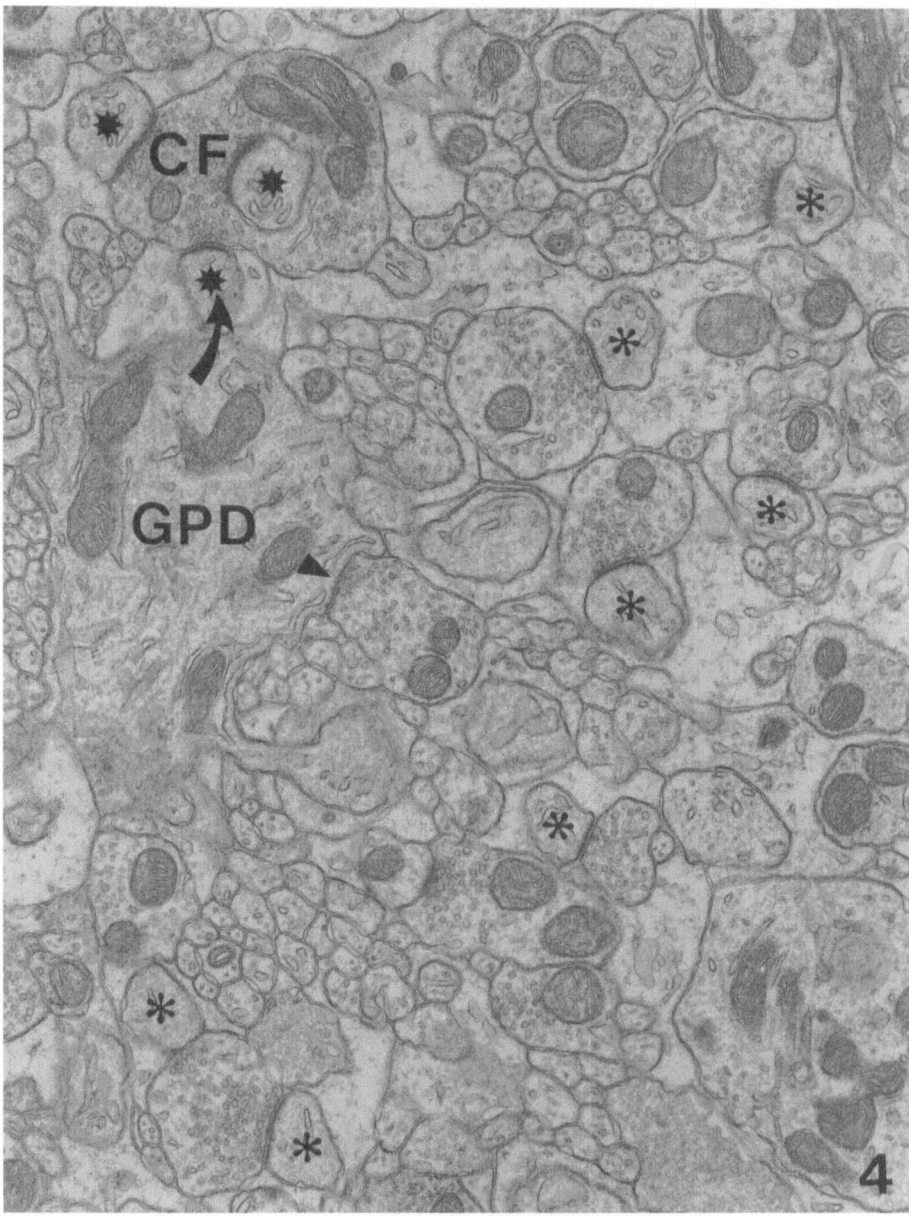

Fig. 4. Electron micrograph of the molecular layer neuropil 2 months after grafting. A thick dendritic profile of a grafted Purkinje cell *(GPD)* is found in the left side of the micrograph. A few dendritic spines emerge from this dendrite; only one of them *(curved arrow)* establishes synaptic contact. The presynaptic bouton exhibits the features of a climbing fiber varicosity *(CF)* and synapses on three spines *(stars)*. An axon terminal, belonging to a stellate cell axon, establishes a symmetrical synapse *(arrowhead)* on the smooth surface of the dendritic profile. In the neuropil, abundant spines *(asterisks)* emerging from distal, grafted Purkinje cell dendrites are synaptically contacted by host parallel fiber varicosities. ×22 000

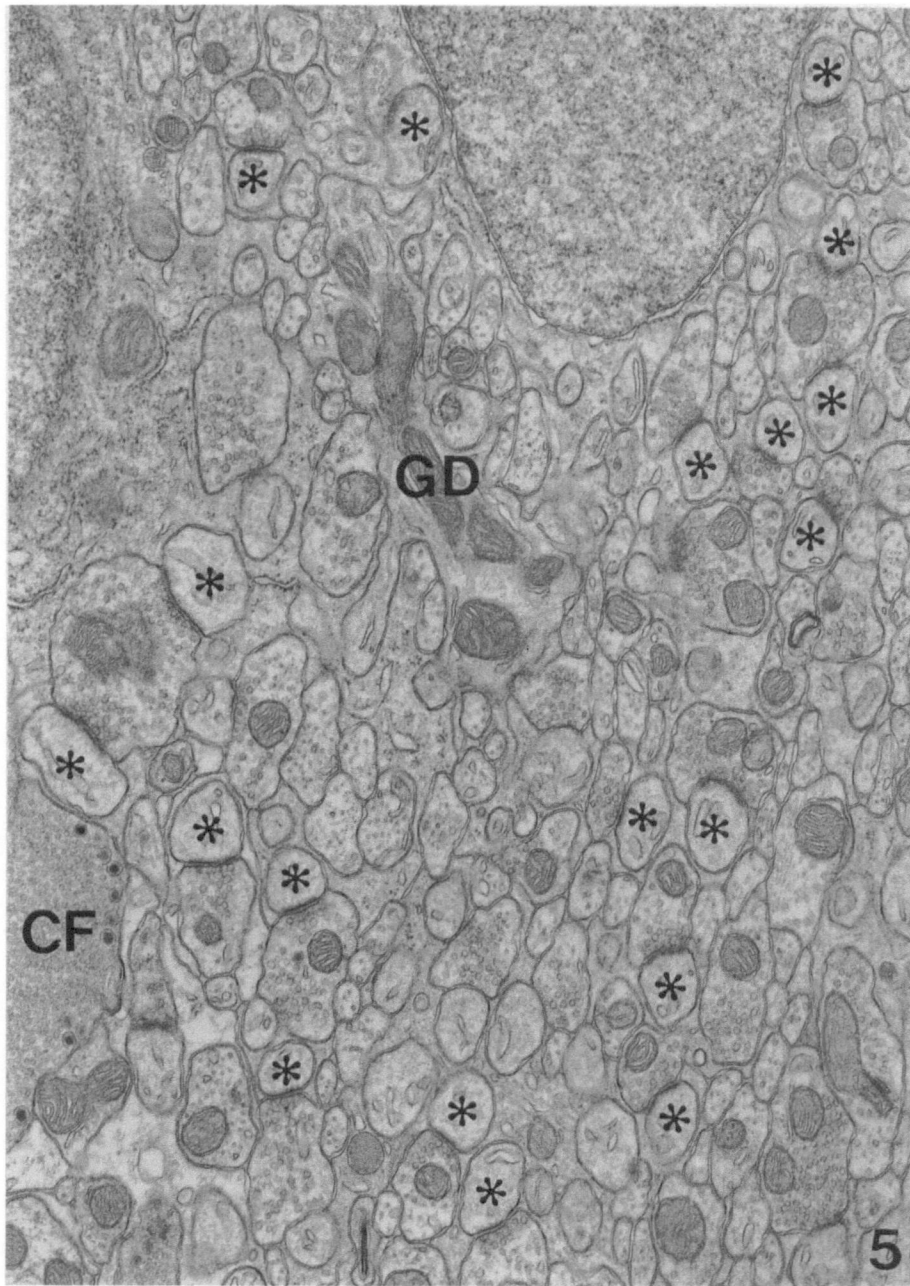

Fig. 5. Electron micrograph illustrating the molecular layer 3 months after grafting. In the neuropil, there is a thin, distal dendritic profile *(GD)* belonging to a grafted Purkinje cell. Note the large amount of spines *(asterisks)* belonging to spiny branchlets that are synaptically contacted by host parallel fibers. *CF* climbing fiber. ×21 000

techniques on *pcd* cerebellar slices in vitro (Gardette et al. 1988). The grafted Purkinje cells were impaled with intracellular microelectrodes and their synaptology was analyzed by electrical stimulation of the white matter at the base of the folium, to anterogradely activate climbing and mossy fibers. All grafted Purkinje cells responded to this white matter stimulation by a typical all-or-none climbing fiber EPSP (complex spike), indicating not only that all grafted Purkinje cells have been synaptically contacted by axons of the host inferior olivary neurons, but also that they are − as in control cerebellum − mono-innervated by climbing fibers. Purkinje cell responses due to the activation of the mossy fiber-granule cell pathway were also elicited in the grafted neurons when the stimulus intensity was low enough not to give rise to any climbing fiber response. The latencies of the disynaptic responses (ranging between 2.4 and 5.2 ms) were longer than those consecutive to the monosynaptic activation of the climbing fibers. These disynaptic EPSPs were markedly graded with the stimulus intensity, as in normal cerebellum. Finally, well-developed IPSPs were also evoked in grafted Purkinje cells, although of shorter duration than in control Purkinje cells. Thus, the electrophysiologic experiments confirm that all the excitatory and inhibitory inputs forming the synaptic investment of grafted neurons are functional, and have characteristics comparable to those in control mice. All these results show that the grafted Purkinje cells have been synaptically integrated into the cortical circuitry of the deficient host cerebellum, fulfilling the second requirement for possible restoration of the mutant cerebellar circuitry.

3. The only output of the cerebellar cortex is the Purkinje cell axonal projection. Thus, the restorative value of neuronal grafting is dependent on the ability of the grafted Purkinje cells to grow axons which, by projecting to proper terminal domains, could establish specific synaptic projections with target neurons, mainly in the deep cerebellar nuclei. Immunocytochemical investigations of Purkinje cell markers have been extremely valuable for studying this problem. When the distance between grafted Purkinje cells and host deep nuclei is only 100−200 μm, numerous immunoreactive axons emerging from molecular layer Purkinje cells reach the deep nuclei through the white matter. In cases in which this distance approaches 700 μm, the immunopositive Purkinje cell axons remain mainly at cortical levels; a few thin bundles reach the most dorsal region of the host deep nuclei, providing a very low density of innervation. Immunoelectron microscopy has permitted identification of the labeled axon varicosities synapsing on deep nuclear neurons. Hence, despite some important restrictions, grafted Purkinje cells seem to be able to send projections to the deep nuclei of the host cerebellum. The distance between the attractive forces (the host nuclear neurons) and the origin of the corticonuclear projection (the grafted Purkinje cells) appears to be a limiting factor for the successful development of corticonuclear interactions. The third prerequisite for neuronal replacement in "point-to-point" systems seems thus to be at least partially satisfied, even though it is the most restrictive.

Conclusions

These studies clearly indicate that Purkinje cell replacement in adult *pcd* cerebellum is, to a certain extent, possible and that this replacement can occur with complete synaptic integration of the grafted neurons into the deficient host cerebellar circuitry. Hence, the results reviewed here provide a solid basis favoring the notion of functional restorative capabilities of neural grafts in systems where, like the cerebellum, the neurons are connected in a precise point-to-point manner. However, the amount of repaired cerebellar circuitry is still too small to consider the possibility of a real improvement of the impaired motor behavior in the transplanted mutant mice.

Furthermore, the grafted Purkinje cells, within the host molecular layer, succeed in acquiring
1. Flattened dendritic trees, spanning the whole molecular layer and composed of a proximal compartment of thick, almost smooth branches, and a distal compartment of spiny branchlets, as normal Purkinje cells do
2. A qualitatively normal synaptic investment.

We can, therefore infer that the development of these neurons follows rules very similar to those regulating normal dendritic differentiation and synaptogenesis (Sotelo 1978). Therefore, adult host neurons confronted with embryonic grafted cells have the ability to establish the required cell-to-cell interactions, leading to the correct maturation and synaptic integration of the grafted neurons (Sotelo and Alvarado-Mallart 1987b).

References

Gardette R, Alvarado-Mallart RM, Crepel F, Sotelo C (1988) Electrophysiological demonstration of a synaptic integration of transplanted Purkinje cells into the cerebellum of the adult Purkinje cell degeneration mutant mouse. Neuroscience 24:777−789

Holmes G (1907) A form of familial degeneration of the cerebellum. Brain 30:466−489

Mullen JR (1977) Site of *pcd* gene action in Purkinje cell mosaicism in cerebellar of chimaeric mice. Nature 270:245−247

Mullen RJ, Eicher EM, Sidman RL (1976) Purkinje cell degeneration, a new neurological mutation in the mouse. Proc Natl Acad Sci USA 73:208−213

Sotelo C (1978) Purkinje cell ontogeny: formation and maintenance of spines. Prog Brain Res 48:149−170

Sotelo C, Alvarado-Mallart RM (1986) Growth and differentiation of cerebellar suspensions transplanted into the adult cerebellum of mice with heredo-degenerative ataxia. Proc Natl Acad Sci USA 83:1135−1139

Sotelo C, Alvarado-Mallar RM (1987a) Reconstruction of the defective cerebellar circuitry in adult Purkinje cell degeneration mutant mice by Purkinje cell replacement through transplantation of solid embryonic implants. Neuroscience 20:1−22

Sotelo C, Alvarado-Mallart RM (1987b) Embryonic and adult neurons interact to allow Purkinje cell replacement in mutant cerebellum. Nature 327:421−423

Wassef M, Sotelo C, Cholley B, Brehier A, Thomasset M (1987) Cerebellar mutations affecting the postnatal survival of Purkinje cells in the mouse disclose a longitudinal pattern of differentially sensitive cells. Dev Biol 124:379−389

Spontaneous Remyelination and Intracerebral Grafting of Myelinating Cells in Mammals

M. Gumpel, O. Gout, and *A. Gansmuller*

Summary

In the early 1960s, remyelination was first observed in animal models of demyelination and described a few years later in multiple sclerosis. These findings were confirmed by a number of experimental works and the remyelination problem is very well documented. The remyelination process depends upon the species and the demyelinating agent; it is more efficient in the case of acute demyelination. Both Schwann cells and oligodendrocytes participate in remyelination. Remyelination by transplantation of myelin-forming cells has been attempted. Transplanted Schwann cells can remyelinate a CNS lesion, but the remyelination is limited to the area of implantation. Little or no migration is observed. Oligodendrocytes or precursor cells are much more invasive and have been shown to migrate from the implantation site to the lesion up to a distance of several millimeters. Thus, remyelination by transplantation of myelin-forming cells is possible, at least in animal models, with the oligodendrocytes being the most efficient. However, it appears that enhancing spontaneous remyelination is a greater challenge and for this purpose, the technique of oligodendrocyte transplantation is probably a very efficient tool.

Introduction

Myelination in the central nervous system (CNS) is normally carried out by oligodendrocytes. Various techniques have been to study oligodendrocyte differentiation and the myelination and remyelination process, including biochemistry, light and ultrastructural morphology, immunology, and in vitro culture (reviews by Morell 1984; Norton 1984).

With the development of the technique of intracerebral transplantation, a new field of investigation was opened. By introducing exogenous glial cells into the brain of a host animal during the normal process of myelination or in the presence of a demyelinated lesion, it became possible to study the behavior of transplanted cells in situ in their normal environment or in pathological stiuations. It became very challenging to obtain a system which would allow researchers to distinguish the transplanted cells themselves, or at least the myelin formed by the transplanted cells, from host myelin. Several models are presently used and considerable new information on myelin-forming cell biology has been gathered. In this paper, we shall briefly review the literature on spontaneous remyelination and remyelination by transplantation of Schwann cells or oligodendrocytes. We also discuss the value of intracerebral transplantation of myelin-forming cells in both therapy and research.

Spontaneous Remyelination in Mammals

Remyelination following a primary demyelination in the CNS is now considered to be the rule rather than the exception. However, this was denied for a long time (Greenfield 1958). The idea of an absence of remyelination stemmed largely from the observation of persistent areas of demyelination in multiple sclerosis. It was challenged by the studies of animal models of demyelination. Bunge et al. (1961) established, for the first time, the existence of spontaneous remyelination in the adult cat spinal cord after demyelination caused by cerebrospinal fluid barbotage. After a stage of complete demyelination, the authors observed the presence of axons ensheathed by thin myelin (the constant ratio axon diameter/myelin sheath thickness normally observed in the adult animal was no longer found). Moreover, the relationships between the remyelinating cells and the newly formed thin myelin sheath were similar to those seen during development (spiral wrapping and compaction). After this pioneer work, immediately followed by that of Bornstein et al. (1962), the development of experimental animal models of demyelination showed that extensive remyelination by oligodendrocytes might occur rapidly in the adult CNS. This was demonstrated in different species after demyelination induced by chemicals: cuprizone diet (Blakemore 1974; Ludwin 1978), lysolecithin (Blakemore 1976; Arenella and Herndon 1984), ethidium bromide (Yajima and Suzuki 1979; Blakemore 1982), 6-aminonicotinamide injections (Blakemore 1975), antisera (EAE; Prineas et al. 1969; Raine 1984), virus infections (JHM hepatitis virus; Herndon et al. 1977), Theiler virus (Dal Canto and Lipton 1975), and herpes virus simplex (Flynn and Martin 1983; Openshaw and Ellis 1984).

The precise mechanisms leading to demyelination are not always well understood. Roughly, the damage can be exerted on the myelin sheath and/or directly on the oligodendrocyte which can be destroyed or modified. It is important to know that the remyelination does not depend on the target of the demyelinating agent; it occurs in the presence or absence of surviving oligodendrocytes in the lesion. Cuprizone diet, which is considered damaging to oligodendrocytes, is followed by extensive spontaneous remyelination (Ludwin 1978), as is a demyelinated lesion due to lysolecithin, whose target is considered to be the myelin sheath (Blakemore 1975). By contrast, the remyelination process appears to be variable according to the species. For instance, after a demyelination obtained by lysolecithin injection, remyelination is almost achieved after 30 days in the rat, while in the rabbit a majority of axons remained persistently demyelinated 6 months after the injection (Blakemore 1978). Moreover, in the same species, remyelination appears to be more efficient in cases of acute rather than chronic demyelination, as shown by Ludwin (1978) after a cuprizone diet in the mouse.

In multiple sclerosis, the occurrence of remyelination was first noted by Périer and Grégoire (1965) and confirmed by many authors (reviewed by Prineas 1985). The newly formed myelin was identified using criteria such as uniformly thin myelin over the whole internodal length and abnormally short internodes (Prineas and Connel 1979). In this human disease, some remyelination occurs at the periphery of chronic lesions but its extent is generally limited (Prineas and Raine 1976; Prineas and Connel 1979; Ludwin 1981; Raine 1983; Harrison 1983). However, Lassmann (1983) described thinly myelinated fibers in fresh lesions and concluded that, in acute forms of the disease, remyelination can begin in acute plaques within a short period of time. This early remyelination could contribute to clinical remission and could even result in the disappearance of the plaque (Prineas 1985). If it is not yet completely demonstrated that clinical remission in multiple sclerosis can be partially explained by spontaneous remyelination, the hypothesis seems reasonable, since Smith et al. (1979), studying an experimental demyelination model, showed that spontaneous oligodendrocyte remyelination restores secure conduction in the rat. In multiple sclerosis the remyelination is considered to be clinically not very efficient; this is probably due to the fact that in subacute forms the remyelination is only effective at the periphery of the plaques and to the possibility of a second wave of demyelination after a first remyelination (Prineas 1985). Why does remyelination not extend and persist in multiple sclerosis? At this point, we touch on an aspect of multiple sclerosis which is still under discussion: what is the primary cause of the disease? If a number of oligodendrocytes survive in acute plaques (Lassmann 1983), one can suppose that these surviving oligodendrocytes themselves are responsible for remyelination since it has been suggested that adult, differentiated oligodendrocytes are able to myelinate again (Wood and Bunge 1986; Szuchet 1987; Lubetzki et al. 1988). Thus, if one hypothesizes that multiple sclerosis is a disease restricted to a population of oligodendrocytes, as suggested by Elias (1987), a second wave of myelination by the same oligodendrocytes cannot be successful. By contrast, if one hypothesizes that the oligodendrocytes are normal in multiple sclerosis, the persistence of demyelinating factors could explain the failure of persistent remyelination.

In the case of remyelination in the CNS, in animal models or in multiple sclerosis, participation of the Schwann cells (peripheral nervous system (PNS) myelinating cells) was very often observed, particularly in the spinal cord, but also in the brain. In animal models of demyelination (lysolecithin injection in the rat), the axons in the center of the lesion were remyelinated by Schwann cells and those of the periphery were remyelinated by oligodendrocytes (Blakemore 1976). Astrocytes were only present in the areas where remyelination by oligodendrocytes occurred. It seems that

1. the astrocytes are required for oligodendrocytes to myelinate, and
2. Schwann cell invasion and myelination is much better in areas where astrocytes have been destroyed (Blakemore 1983).

The Schwann cells remyelinating spontaneously in the spinal cord probably come from the peripheral roots. In the brain, Blakemore (1983) hypothesized that they originate from blood vessel innervation. One of the limiting factors for the spontaneous remyelination of Schwann cell in the CNS is the presence of the glial limiting membrane, which covers not only the subglial surfaces but also the surfaces apposed to blood vessels (Blakemore 1983). However, this glial limiting membrane is interrupted in traumatic experiments (injections of chemicals) or in pathological situations such as virus infection and multiple sclerosis. In this latter disease, it is now well established that Schwann cells can invade plaques, mostly in the spinal cord (50% of cases according to Itoyama et al. 1980) and form peripheral-type myelin around demyelinated central axons (Feigin and Popoff 1966; Feigin and Ogata 1971; Ghatak et al. 1973; Itoyama et al. 1983). To our knowledge, it is not yet established if remyelination of central axons by Schwann cells restores normal conduction.

Remyelination in Adult Mammals by Schwann Cell Transplantation

A first attempt to transplant Schwann cells into the spinal cord to remyelinate a demyelinated lesion was published by Blakemore (1977). The author created a lesion by injection of lysolecithin to the spinal cords of rats and cats. In this situation, the lesion is normally spontaneously remyelinated by oligodendrocytes and Schwann cells. In order to avoid this spontaneous repair, the animals were subjected to X-ray irradiation, which completely suppressed both oligodendrocyte and Schwann cell remyelination. A piece of dissociated sciatic nerve was placed over the injected area to provide a source of viable Schwann cells. Under these conditions, the demyelinated lesion was partially remyelinated by transplanted Schwann cells. During the following years, Blakemore used the same technique (chemical injection and irradiation) to obtain areas of persistent demyelination and established that Schwann cells, in various conditions (even purified in culture), were able to remyelinate central demyelinated axons (Blakemore 1980, 1984; Blakemore and Crang 1985). However, the remyelination was always limited to the area of transplantation and to the periphery of the blood vessels.

In 1981, Duncan et al. transplanted cultured Schwann cells to remyelinate lesions induced by lysolecithin in the mouse spinal cord. To confirm the remyelination by transplanted Schwann cells, an elegant procedure was used. The authors transplanted rat Schwann cells into mouse spinal cord (xenograft). The hosts were treated by antilymphocytic serum (ALS). After several weeks, the ALS was stopped and the animals received an immune cell transfer. In these conditions, the xenografted Schwann cells were rejected and this proved definitively that myelinating cells were exogenous. However, the authors concluded that cell remyelination was limited to the adjacent site of the original implant. There was little or no evidence of Schwann cell migration.

Remyelination in the Adult Mammal by Oligodendrocyte Transplantation

Peripheral myelin can be easily distinguished from central myelin because of the presence of a basal lamina around the myelin sheath or by immunohistochemistry (presence of P_0 specific to PNS myelin). The only difficulty was to distinguish

Schwann cell spontaneous remyelination from transplanted Schwann cell remyelination. Blakemore (1977) resolved the problem by creating reliable persistent demyelinated areas, and Duncan et al. (1981) did so by showing rejection of xenografted Schwann cells.

In the case of transplantation of oligodendrocytes the major difficulty was to distinguish myelin synthesized by transplanted cells from that formed by the host CNS cells. We resolved this difficulty by using the mutant shiverer mouse as host. The shiverer is an autosomic recessive mutation due to the deletion of 5 of 7 axons of the gene coding for myelin basic protein (MBP) on chromosome 18 (Roach et al. 1985; Sidman et al. 1985). The homozygous animals (shi/shi) are deprived of biochemically detectable MBP (Dupouey et al. 1979). This defect is correlated with the absence of major dense line (MDL) at ultrastructural observation of the myelin sheath (Privat et al. 1979). Using this model, we could study the behavior of normal newborn oligodendrocytes or precursor cells transplanted into the newborn shiverer brain and thus integrated into the brain during the normal process of myelination. These oligodendrocytes were shown to be able to survive in the host brain, to migrate over long distances during the myelination process, and to myelinate host axons. This was demonstrated by the presence of patches of MBP/MDL-positive myelin from the graft all over the host brain (Gumpel et al. 1983, 1987; Lachapelle et al. 1984; Gansmuller et al. 1986; Baulac et al. 1987). The migration of oligodendrocyte precursor cells has also been demonstrated by Small et al. (1987). During the development of the rat, the optic nerve is progressively invaded by O_2A (oligodendrocyte-astrocyte 2; Raff et al. 1983) precursor cells from the chiasmal end to the retina. This suggests that, from their origin in the germinal epithelium, the precursors of myelin-forming cells sometimes have to migrate a long way to reach their definitive myelination area. Thus, at a stage of their differentiation they are endowed with very important migrative properties.

There are several models allowing the detection of the myelin formed by the transplanted cell based on histochemical and/or ultrastructural characteristics. However, in the experiment we describe in this paper, we used the so-called shiverer model.

The first step of our experiment was to induce a demyelinated area in an adult shiverer host. The demyelinated lesion (Fig. 1) in the spinal cord of a 1.5- to 2-month-old shiverer mouse was made by injection of 2 μl of 1% lysolecithin (lysophosphatidyl choline; Sigma) in saline. The site of injection was marked with charcoal. We observed that such a lesion is spontaneously remyelinated by host cells (Fig. 2), the remyelinated axons being ensheathed by typically shiverer myelin (without MDL). The chronology of events is comparable to that of remyelination of the same lesion in the normal mouse (unpublished results). In an experimental series, a fragment of newborn normal mouse containing oligodendrocytes and precursor cells was grafted simultaneously in the spinal cord of the demyelinated animals at a distance of one to three intervertebral segments (3 to 8 mm), rostral or caudal to the lesion (Gout et al. 1988a, b). In these experimental conditions, it was clearly observed that axons of the demyelinated lesion were remyelinated by both grafted (MDL-positive myelin) and host (typical shiverer myelin) oligodendrocytes (Fig. 3). This means that oligodendrocytes or precursor cells contained in the graft were able to migrate in the host tissues and to myelinate

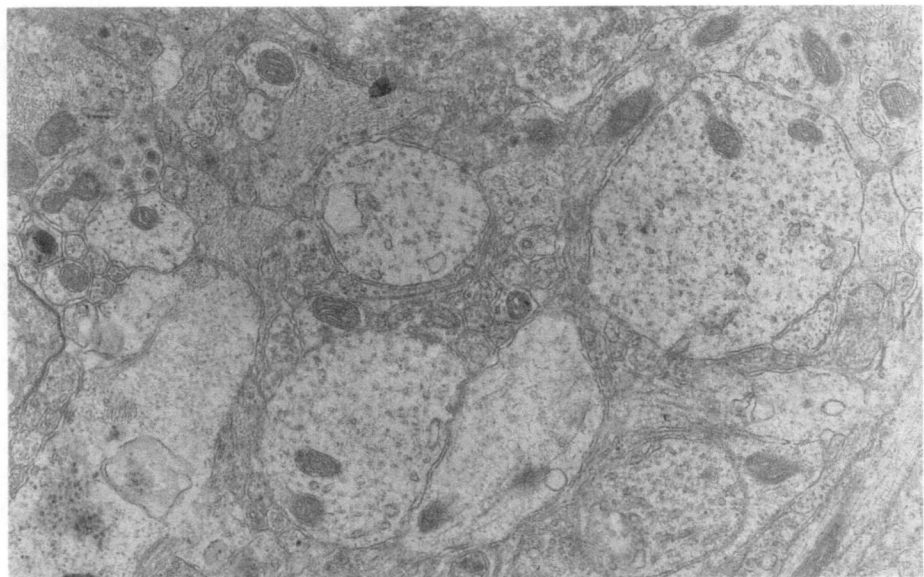

Fig. 1. Demyelination in the spinal cord of an adult shiverer mouse, 7 days after injection of lysolecithin (× 16 000)

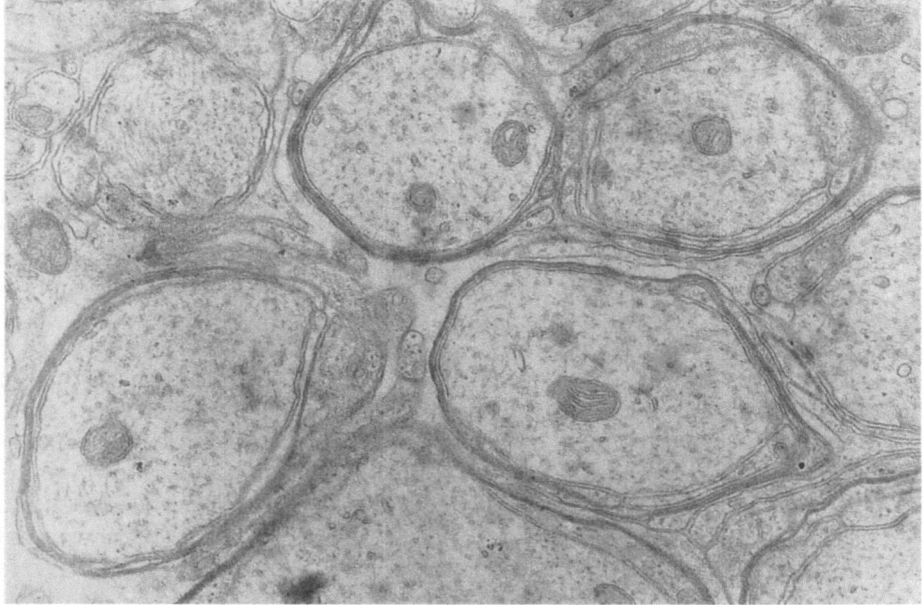

Fig. 2. Spontaneous remyelination in the spinal cord of a shiverer mouse 76 days after injection of lysolecithin. The newly formed myelin is shiverer-type myelin (× 16 000)

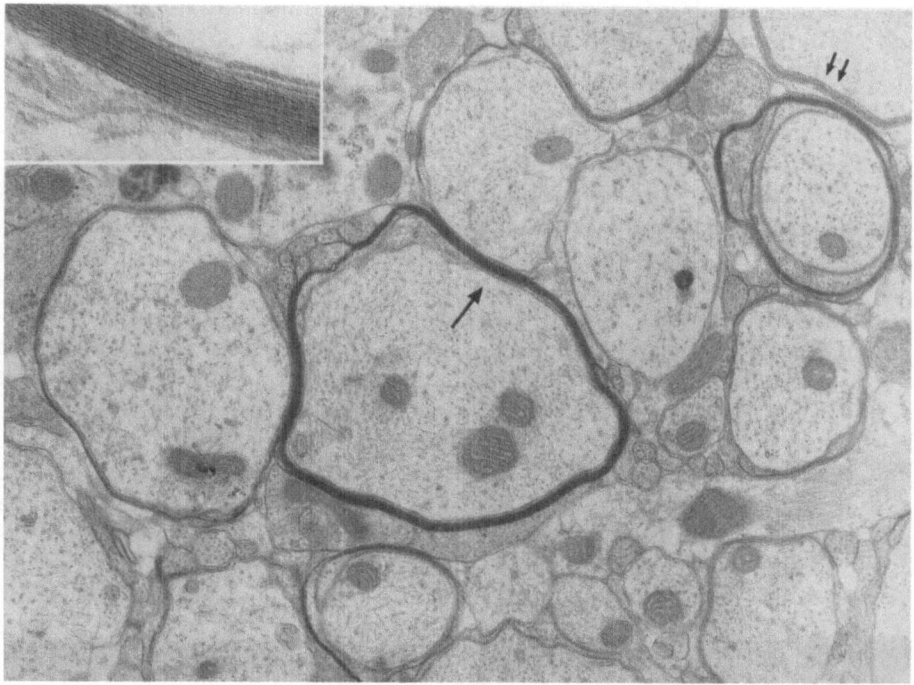

Fig. 3. Remyelination in a lesion induced by lysolecithin in the spinal cord of an adult shiverer mouse. A fragment of newborn CNS tissue containing oligodendrocytes and precursor cells was grafted 6 mm from the lesion: 94 days after the operation, the demyelinated axons are remyelinated by both shiverer (⇉) and normal (→) myelin (× 16 000). *Inset:* high magnification of the normal myelin with MDL (× 80 000)

demyelinated axons. As suggested by Blakemore (1983), this experiment shows that nude axons express a signal which can be received by cells even at a distance of 8 mm. These cells are then able to migrate to the lesion through the adult tissues. Blakemore and Crang (1988), using the model described above of persistent demyelinating area (chemical injection and irradiation), observed that cultured CNS cells containing oligodendrocytes transplanted into the lesion were able to remyelinate very efficiently. The authors have no specific marker for transplanted oligodendrocytes or myelin formed by these cells but they are confident enough in this persistent demyelination model to affirm that the transplanted oligodendrocytes are responsible for the remyelination observed.

Conclusions

Remyelination of demyelinated lesion is possible by transplantation of Schwann cells and oligodendrocytes. This means that both types of myelin-forming cells survive transplantation in an adult brain, can be integrated into the environment, and synthesize myelin around the host demyelinated axons. Oligodendrocytes can be considered more efficient than Schwann cells for several reasons:

1. They invade the host parenchyma extensively and are capable of migration, while the Schwann cells remain restricted to the site of injection and remyelinate only around the blood vessels.

2. The spontaneous remyelination by oligodendrocytes restores a secure conduction. This is probably true for transplanted oligodendrocytes. Nothing is known about central conduction after remyelination by Schwann cells. It should be stressed, however, that the chemical composition of peripheral myelin is slightly different from that of CNS myelin. The wrapping of the axon is also different, with the oligodendrocyte ensheathing several axons and the Schwann cell ensheathing only one. There is a basal lamina around the PNS myelin, but not around the CNS myelin, and the node of Ranvier is organized differently.

3. Until recently, it was possible to purify Schwann cells but not oligodendrocytes. Nowadays, the techniques to obtain pure oligodendrocytes and even pure progenitor cells are improving. Thus, it will soon be possible to transplant pure myelin-forming cells of both types. However, from the therapeutic point of view, is it necessary to transplant exogenous myelin-forming cells to remyelinate a lesion, since it has now been proven that, even in multiple sclerosis, these cells can be recruited in the brain itself (see review Prineas 1985)? In the case of multiple sclerosis, the problem, at least in acute plaques, does not seem to be the absence of remyelination but the destruction of successive waves of remyelination because of the persistence of the disease. In this view, the transplantation of oligodendrocytes could eventually be valuable as a transient aid for remyelination. Moreover, in the hypothesis of a defect of a restricted population of oligodendrocytes, transplantation could be useful to locally stop the disease. But this hypothesis is far from being confirmed. For the moment it appears to us much more challenging to try to understand which factors can inhibit or enhance spontaneous remyelination. However, the technique of intracerebral transplantation could be a very efficient tool to study which cells of the CNS are recruited for remyelination − precursor quiescent cells? mature oligodendrocytes? − and what factors are favorable to their recruitment and migration to the lesion.

References

Arenella WF, Herndon RM (1984) Mature oligodendrocyte division following experimental demyelination in adult animals. Arch Neurol 41:1162−1165

Baulac M, Lachapelle F, Gout O, Berger B, Baumann N, Gumpel M (1987) Transplantations of oligodendrocytes in the newborn mouse brain: extension of myelination by transplanted cells. Anatomical study. Brain Res 420:39−47

Blakemore WF (1974) Remyelination of the superior cerebellar peduncle in old mice following demyelination induced by cuprizone. J Neurol Sci 22:121−126

Blakemore WF (1975) Remyelination by Schwann cells of axons demyelinated by intraspinal injection of 6-aminonicotinamide. J Neurocytol 4:745−757

Blakemore WF (1976) Invasion of Schwann cells into the spinal cord of the rat following local injections of lysolecithin. Neuropathol Appl Neurobiol 2:21–39

Blakemore WF (1977) Remyelination of CNS axons by Schwann cells transplanted from the sciatic nerve. Nature 266:68–69

Blakemore WF (1978) Observations on remyelination in the rabbit spinal cord following demyelination induced by lysolecithin. Neuropathol Appl Neurobiol 4:47–69

Blakemore WF (1980) The effect of subdural nerve tissue transplantation on the spinal cord of the rat. J Neuropathol Appl Neurobiol 6:433–447

Blakemore WF (1982) Ethidium bromide induced demyelination in the spinal cord of the cat. Neuropathol Appl Neurobiol 8:365–375

Blakemore WF (1983) Remyelination of demyelinated spinal cord axons by Schwann cells. In: Kao CC, Bunge RP, Reier PJ (eds) Spinal cord reconstruction. Raven, New York, pp 281–291

Blakemore WF (1984) Limited remyelination of CNS axons by Schwann cells transplanted into the subarachnoid space. J Neurol Sci 64:265–276

Blakemore WF, Crang AJ (1985) The use of cultured autologous Schwann cells to remyelinate areas of persistent demyelination in the central nervous system. J Neurol Sci 70:207–223

Blakemore WF, Crang AJ (1988) Extensive oligodendrocyte remyelination following injection of cultured central nervous system cells into demyelinating lesions in adult central nervous system. Dev Neurosci 10:1–11

Bornstein MB, Appel SH, Murray MR (1962) The application of tissue culture to the study of experimental "allergic" encephalomyelitis. Demyelination and remyelination. In: Jacob H (ed) Proceedings of the 4th International Congress of Neuropathology, vol 2. Thieme, Stuttgart, pp 279–282

Bunge MB, Bunge RB, Ris H (1961) Ultrastructural study of remyelination in adult rat spinal cord. J Biophys Biochem Cytol 10:67–94

Dal Canto MC, Lipton HL (1975) Primary demyelination in Theiler's virus infection: an ultrastructural study. Lab Invest 33:626–637

Duncan ID, Aguayo AJ, Bunge RP, Wood PK (1981) Transplantation of rat Schwann cells grown in tissue culture into the mouse spinal cord. J Neurol Sci 49:241–252

Dupouey P, Jacque C, Bourre JM, Cesselin F, Privat A, Baumann N (1979) Immunochemical studies of myelin basic protein in shiverer mouse devoid of major dense line of myelin. Neurosci Lett 12:113–118

Elias SB (1987) Oligodendrocyte development and the natural history of multiple sclerosis. Arch Neurol 44:1294–1299

Feigin I, Ogata J (1971) Schwann cells and peripheral myelin within human central nervous tissue: the mesenchymal character of Schwann cells. J Neuropathol Exp Neurol 30:603–611

Feigin I, Popoff N (1966) Regeneration of myelin in multiple sclerosis: the role of mesenchymal cells in such regeneration and in myelin formation in the peripheral nervous system. Neurology (Minneap) 16:364–372

Flynn TE, Martin JR (1983) Topography of remyelinated chronic spinal cord lesions in herpes simplex virus type 2 infections of mice. J Neurol Sci 61:327–339

Gansmuller A, Lachapelle F, Baron-Van Evercooren A, Hauw JJ, Baumann N, Gumpel M (1986) Transplantations of newborn CNS fragments into the brain of shiverer mutant mice: extensive myelination by transplanted oligodendrocytes. II. Electron microscopic study. Dev Neurosci 8:197–207

Ghatak NR, Hirano A, Doron Y, Zimmerman HM (1973) Remyelination in multiple sclerosis with peripheral type myelin. Arch Neurol 29:262–267

Gout O, Gansmuller A, Baumann N, Gumpel M (1988a) Remyelination of demyelinated lesion in the adult shiverer spinal cord by transplantation of normal neonatal cells. In: Confavreux C, Aimard G, Devic M (eds) Trends in European multiple sclerosis research. Elsevier, Amsterdam, pp 115–119

Gout O, Gansmuller A, Baumann N, Gumpel M (1988b) Remyelination by transplanted oligodendrocytes of a demyelinated lesion in the spinal cord of the adult shiverer mouse. Neurosci Lett 87:195–199

Greenfield JG (1958) Demyelinating diseases. In: Greenfield JG, Blackwood W, McMenemey WH, Meyer A, Norman RM (eds) Neuropathology. Arnold, London, pp 441–474

Gumpel M, Baumann N, Raoul M, Jacque C (1983) Survival and differentiation of oligoden-drocytes from neural tissues transplanted into newborn mouse brain. Neurosci Lett 37:307–311

Gumpel M, Lachapelle F, Gansmuller A, Baulac M, Baron-Van Evercooren A, Baumann N (1987) Transplantation of human oligodendrocytes into the shiverer brain. Ann NY Acad Sci 495:71–85

Harrison BM (1983) Remyelination in the cerebral nervous system. In: Hallpike JF, Adams CW, Tourtellotte WW (eds) Multiple sclerosis. Williams and Wilkins, Baltimore, pp 461–478

Herndon RM, Price DL, Weiner LP (1977) Regeneration of oligodendroglia during recovery from demyelinating disease. Science 195:693–694

Itoyama Y, Webster H de F, Richardson EP Jr, Trapp BD (1980) Schwann cell remyelination of demyelinated axons in spinal cord multiple sclerosis lesions. Ann Neurol 71:167–177

Lachapelle F, Gumpel M, Baulac M, Jacque C, Duc P, Baumann N (1984) Transplantation of CNS fragments into the brain of shiverer mutant mice: extensive myelination by implanted oligodendrocytes. Immunohistochemical studies. Dev Neurosci 6:325–334

Lassmann H (1983) Comparative neuropathology of chronic experimental allergic encephalomyelitis and multiple sclerosis. Springer, Berlin Heidelberg New York

Lubetzki C, Gansmuller A, Lachapelle F, Lombrail P, Gumpel M (1988) Myelination by oligodendrocytes isolated from 4 weeks old rat brain CNS and transplanted into the newborn shiverer brain. J Neurol Sci 88:161–175

Ludwin SK (1978) Central nervous system demyelination and remyelination in the mouse. An ultrastructural study of cuprizone toxicity. Lab Invest 39:597–612'

Ludwin SK (1980) Chronic demyelination inhibits remyelination in the central nervous system: an analysis of contributing factors. Lab Invest 43:382–387

Ludwin SK (1981) Pathology of demyelination and remyelination. In: Waxman SG, Ritchie JM (eds) Demyelinating disease: basic and clinical electrophysiology. Raven, New York, pp 123–168

Morell P (1984) Myelin, 2nd edn. Plenum, New York

Norton WT (1984) Oligodendroglia. In: Agranoff BW, Aprison MH (eds) Advances in neurochemistry, vol 5. Plenum, New York

Openshaw H, Ellis W (1984) Demyelination and remyelination in the optic nerve of mice infected with herpes simplex virus (HSV). Neurology (NY) [Suppl 1] 34:96

Périer O, Grégoire A (1965) Electron microscopic features of multiple sclerosis lesions. Brain 88:937–952

Prineas JW (1985) The neuropathology of multiple sclerosis. In: Koetsier JC (ed) Demyelinat-ing diseases. Elsevier, Amsterdam, pp 213–225 (Handbook of clinical neurology, vol 3/47)

Prineas JW, Connell FW (1979) Remyelination in multiple sclerosis. Ann Neurol 5:22–31

Prineas JW, Raine CS (1976) Electron microscopy and immunoperoxidase studies of early mul-tiple sclerosis lesions. Neurology (Minneap) 26:29–32

Prineas J, Raine CS, Wisniewski H (1969) An ultrastructural study of experimental demyelina-tion. III. Chronic experimental allergic encephalomyelitis in the central nervous system. Lab Invest 21:472–483

Privat A, Jacque C, Bourre JM, Dupouey P, Baumann N (1979) Absence of the major dense line in the mutant mouse shiverer. Neurosci Lett 12:107–112

Raff MC, Miller RH, Noble M (1983) A glial progenitor cell that develops in vitro into an astro-cyte or an oligodendrocyte depending on culture medium. Nature 303:390–396

Raine CS (1983) Multiple sclerosis and chronic relapsing EAE: comparative ultrastructural neuropathology. In: Hallpike SF, Adams CWM, Tourtellotte WW (eds) Multiple sclerosis: Pathology, diagnosis and management. Williams and Wilkins, Baltimore, pp 413–460

Raine CS (1984) Biology of disease. Analysis of autoimmune demyelination: its impact upon multiple sclerosis. Lab Invest 50:608–635

Roach A, Takahashi N, Pravtcheva D, Ruddle F, Hood L (1985) Chromosomal mapping of mouse myelin basic protein gene and structure and transcription of the partially deleted gene in shiverer mutant mice. Cell 42:149–155

Sidman RL, Conover CS, Carson JH (1985) Shiverer gene maps near the distal end of chromo-some 18 in the mouse. Cell Genet 39:241–245

Small RK, Riddle P, Noble M (1987) Evidence for migration of oligodendrocyte-type 2 astro-
cyte progenitor cells into the developing rat optic nerve. Nature 328:155–157
Smith KJ, Blakemore WF, McDonald WI (1979) Central remyelination restores secure conduction. Nature 280:395–396
Szuchet S (1987) Myelin palingenesis: the reformation of myelin by mature oligodendrocytes in
the absence of neurons. In: Althaus HH, Seifert W (eds) Glial-neuronal communication in
development and regeneration. Springer, Berlin Heidelberg New York, pp 755–777
(NATO ASI series)
Wood PM, Bunge RP (1986) Myelination of cultured dorsal root ganglion neurons by oligodendrocytes obtained from adult rat. J Neurol Sci 74:153–169
Yajima K, Suzuki K (1979) Demyelination and remyelination in the rat central nervous system
following ethidium bromide injection. Lab Invest 41:385–392

Synaptic Plasticity vs Pathology in Alzheimer's Disease: Implications for Transplantation

C. W. Cotman

Summary

It is now clear that the brain has a repertoire of plasticity and regenerative mechanisms that may help maintain function in the normal aged brain and perhaps also in the brains of patients with Alzheimer's disease (AD). Some of these same mechanisms (e.g., axon sprouting) may, however, also participate in the development of AD pathologies such as senile or neuritic plaque formation. Accordingly neuronal transplants in the AD brain will experience a complex environment of growth and degeneration which may greatly compromise their prospects for mediating long-term functional recovery.

Introduction

Neural grafts offer great promise for replacing neurons lost in the course of age-related neurodegenerative disorders. A disease such as Alzheimer's disease (AD) presents a particularly challenging situation. First, the cells must survive in an environment undergoing extensive degeneration. Second, they must send out processes, locate the correct target, and compete with reactive sprouting from local neurons. Third, transplanted neurons must receive the correct input and respond properly to it. In all these stages it is necessary for the grafted neurons to avoid the pathology associated with AD, e.g., neuronal degeneration, neurite (senile) plaques and neurofibrillary tangles.

In this article I will examine current information on synaptic growth and plasticity in the AD brain which bears on the prospect of transplantation in AD. Are endogenous AD neurons capable of growth and sprouting? Is it likely the AD brain will recruit new cells into AD pathology? What molecular mechanisms are operative in the AD brain that may alter the course of events? Answers to these and related issues may help design appropriate transplant paradigms and optimize the chances of success. It is, of course, recognized that transplants may have limited utility in AD. Transplants will perhaps be less successful in late-stage AD when degeneration is massive and general. In early-stage AD, however, transplanted cells may be able to replace lost circuitry components and stabilize the brain in the face of progressive transneuronal influences which could contribute to cascades of degeneration.

In the AD brain, complex cortical networks degenerate, placing stringent requirements on transplant integration. One such circuit is that of the entorhinal cortex which projects to the hippocampus. Studies on brain tissues obtained post-mortem from patients with AD have demonstrated severe cell loss in select areas of the limbic system (Kemper 1978). Van Hoesen and colleagues (Hyman et al. 1984, 1986) have reported that AD is accompanied by a selective loss of the layer II stellate cells of the entorhinal cortex and the pyramidal neurons of the sub-iculum. In essence, their findings suggest that the early stages of the disease are characterized by a very selective deafferentation of a brain region critical to mem-ory storage and the major source of input to the hippocampal formation. These lesions are further compounded by direct loss of neurons in the hippocampus itself (Ball et al. 1985). It is becoming clear that damage to entorhinal/hippocampal/sub-icular circuits is a key pathology in AD. In fact, it has been suggested that the ear-liest neuropathology to appear in AD is found in these regions (Hyman et al. 1986).

The AD Brain is Capable of Axon Sprouting

In order for transplants to properly integrate, it is necessary that they become innervated and locate and connect to their target. This requires that AD neurons be capable of axonal growth. A series of studies on the effect of entorhinal lesions in the rodent brain suggests that these lesions release compensatory regrowth by the remaining circuitry. After unilateral ablation of the entorhinal cortex, over 85% of the synapses in the outer two-thirds of the ipsilateral dentate gyrus molecular layer degenerate, representing nearly 60% of the total synaptic input to these neurons (Lee et al. 1977; Matthews et al. 1976a, b). Normal synapses rapidly disappear while degenerating fibers accumulate in the denervated zone. The majority of degenerating axon terminals are removed within 1 week post-lesion, although the clearance of all degeneration products may take several weeks.

New synapses begin to reinnervate targets within 2 to 4 days after entorhinal damage. The maximum rate of synaptic replacement occurs between 10 and 30 days postlesion. Synaptic density is restored to prelesion levels within a few weeks following the lesion (Hoff et al. 1982; Lee et al. 1977; Matthews et al. 1976b). The reinnervated neuropil is morphologically very similar to controls (Matthews et al. 1976b), except for the presence of a few giant boutons which make multiple synap-tic junctions (Hoff et al. 1982).

The cholinergic projection from the septal nuclei is one of the inputs which sprouts in the denervated zone (Cotman et al. 1973; Lynch et al. 1972; Seress and Ribak 1984; Fig. 1). These cholinergic afferents can be easily monitored with histochemical staining for the enzyme acetylcholinesterase (AChE; Lynch et al. 1972). In the normal brain, the dentate gyrus molecular layer shows a light, evenly deposited AChE stain. In marked contrast, an animal that has received an ento-rhinal lesion shows a very dark, intense staining deposit in the outer molecular layer, indicative of the proliferation of the septohippocampal projection. Direct biochemical measurements on AChE and choline acetyltransferase activity support the contention that the septal input to the molecular layer has increased (Nadler et al. 1973).

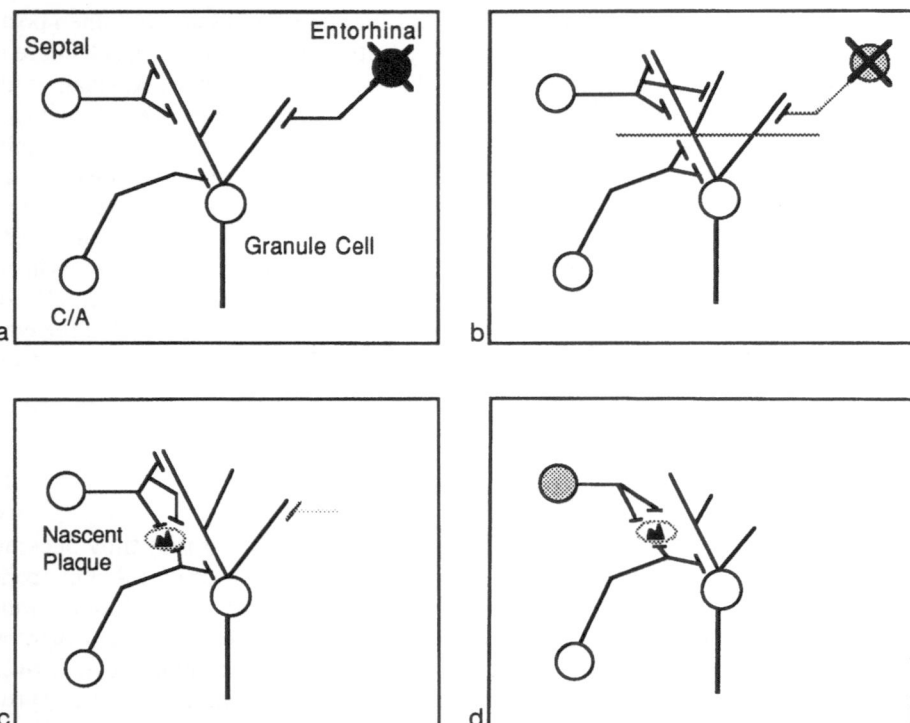

Fig. 1a–d. Model for axon sprouting and senile (neuritic) plaque formation in AD. **a** Partial loss of input releases growth of remaining inputs. Cholinergic septal and commissural associational (C/A) fibers sprout onto dentate granule cells when entorhinal neurons degenerate. In the normal rodent brain these changes are stable and persist a lifetime. In AD, however, a sequence of events may lead to senile plaque formation. **b** Aberrant sprouting occurs in localized regions contributing to the development of plaques. It is likely that-there is an increase in injury-induced neurotrophic factors in the terminal fields of degenerating neurons. This, combined with a suitable localized substrate, provides a stimulus for growth. **c** The plaque grows, fed by a cycle of regenerative growth and degeneration. Neurites become progressively attracted into the plaque. Reduced protease inhibitor activity may promote neurite retraction in normal target areas (e.g., dentate granule cells) vs plaques(?). **d** The progressive accumulation of neurites in the area of the plaque causes an imbalance in normal trophic requirement. This causes some atrophy in target cells and a progressive loss in trophic support for input neurons. These neurons then become more vulnerable to insults, enhancing the probability of degeneration (e.g., viral infection, excitotoxic injury, etc.)

Since entorhinal neurons degenerate in AD, it would be expected that septal neurons sprout in AD. In those AD patients where significant cholinergic input to the hippocampus is present (which is the case in most AD patients), intensification of AChE activity in the outer half of the dentate molecular layer has been observed (Geddes et al. 1985, 1986; Hyman et al. 1987). These results indicate that cholinergic neurons in AD patients, when present, are capable of a sprouting response.

This demonstrates that axonal growth can occur in the AD brain and that the essential mechanisms are still operative. Thus, one of the requirements for successful transplantation is realized. Local sprouting, however, may in itself compete with transplant integration.

Transplants Can Partially Compete with Endogenous Sprouting Caused by Partial Denervation

Transplanted neurons must not only survive and innervate the proper target but, in order to completely restore proper circuitry, they must prevent or reverse heterotypic sprouting by native neurons. For example, in AD it would be appropriate to graft entorhinal neurons to replace those which have degenerated. In order to completely restore cirucitry, it would be necessary to reverse conditions such as cholinergic sprouting. In previous experiments with entorhinal transplants (Gibbs et al. 1985), we demonstrated that entorhinal neurons would reinnervate the dentate gyrus and form a topographically normal input. While in most transplants (34 of 40) the entorhinal projection did not successfully compete with the septal sprouting, in a few cases the AChE intensification appeared to be at least partially reduced. The most successful cases appeared to be associated with the strongest innervation. Thus, the normal input can compete with sprouting and restore circuitry, at least in select cases.

Growth and Sprouting Reactions in the AD Brain Become Associated with AD Pathology

In several cases of AD, we observed that there are numerous AChE-positive plaques in the denervated dentate molecular layer, precisely corresponding to the enhanced sprouting response in this region (Geddes et al. 1986) Their disposition corresponds to the density and distribution of plaques, as revealed by silver stain procedures. Based on evidence obtained to date with available cases, it seems to us that septal sprouting precedes plaque formation. For example, sprouting in the molecular layer exists without plaques, while in adjacent areas of the layer sprouting is seen with plaques as if sprouting progresses to plaque formation. This is consistent with the hypothesis that an early phase of plaque formation is an aberrant process of increased sprouting. The growth of septal fibers may become overstimulated, thereby contributing to an overabundance of cholinergic sprouts within the plaque (Fig. 1).

It could be that sprouting is caused by localized degeneration which causes an increase in trophic factor activity such as we have previously demonstrated is associated with entorhinal lesions (see Cotman and Nieto-Sampedro 1984; Geddes et al. 1986). In the case of AD, however, a local excess of trophic activity may exaggerate the injury-induced response. Reactive astrocytes, and possibly neurite-stimulating cells in the area, may cause enhanced sprouting in local regions feeding the developing plaque. Recent findings, for example, indicate that astrocytes and possibly even macrophages produce growth factors, as well as cell surface molecules such as laminin which can stimulate neurite sprouting of central neurons. A cycle may develop. Septal cells with their input increasingly in plaques lose their normal target cells in the dentate gyrus. Lacking their proper target, the cells are deprived of trophic support, as if they have been axotomized. They then degenerate or at least become more vulnerable to degeneration. This causes a further increase in neurotrophic factors, further sprouting, enhanced plaque formation, and further degeneration.

The possibility of plaque formation as an end point of sprouting needs to be considered as a possible complication of transplantation. Transplant fibers might actually become involved in plaques.

Abnormalities in Endogenous Brain Protease Inhibitors May Contribute to AD Pathology and Influence Restorative Growth

Plaques contain a central amyloid core surrounded by degenerating and sprouting neurites and reactive glia. The recent findings that brain amyloid deposits contain α-1-antichymotrypsin (Abraham et al. 1988) and that the amyloid precursor protein contains a trypsin inhibitor sequence (Ponte et al. 1988; Tanzi et al. 1988; Kitaguchi et al. 1988) indicate that an imbalance of proteases and protease inhibitors may be involved in the etiology of the disease.

Recently we have begun to examine the state of protease nexins in AD (Wagner et al. 1988). Protease nexin-1 (PN-1) is a 44 kDa protease inhibitor synthesized and secreted by a variety of cultured cells including astrocytes (Baker et al. 1980, 1986; Rosenblatt et al. 1987). PN-1 inhibits thrombin, urokinase, and plasmin by forming stable complexes with the catalytic site of the protease (Scott et al. 1985). These PN-1-protease complexes bind to the cells that produce PN-1 and are rapidly internalized and degraded (Low et al. 1981). When PN-1 binds to the extracellular matrix (Farrell et al. 1988), this accelerates its inhibition of thrombin (Farrell and Cunningham 1986) and greatly retards its inhibition of urokinase and plasmin (Wagner et al. 1989a). Thus, PN-1 is relatively specific for thrombin. PN-1 appears to be identical to a glial-derived nexin (Gloor et al. 1986) which stimulates neurite outgrowth in neuroblastoma cells (Monard et al. 1983) and primary sympathetic neurons (Zurn et al. 1988). Previously it has been shown that thrombin causes neurite retraction in neuroblastoma cells (Gurwitz and Cunningham 1988), and that the neurite outgrowth activity of PN-1 depends on its ability to inhibit thrombin or a thrombin-like protease (Monard et al. 1983; Gurwitz and Cunningham 1988).

The activity of PN-1 appears markedly reduced in the AD brain. PN-1 was measured by its ability to form SDS-resistant complexes with ^{125}I-thrombin. ^{125}I-thrombin-protease inhibitor complexes are much lower in the AD brain compared with control brains, indicating a reduced PN-1 activity in AD. PN-1 activity in the AD brain is only 13% of control values (control, 65.5 ± 26.1 pmol/mg protein, $n = 5$; AD, 8.24 ± 5.98 pmol/mg protein, $n = 8$). The reduced PN-1 activity was seen in all of the AD brain samples compared with controls. In fact, the highest AD sample is still about twofold lower than the lowest control value. No significant correlation was found between the decreased PN-1 levels in the AD samples and age, sex, or postmortem delay. Furthermore, the activity of protease nexin-2 (Van Nostrand and Cunningham 1987), which is also synthesized by glial cells, does not vary significantly among these samples.

It appears that the decreased PN-1 activity in the AD brain samples is not due to the presence of an inhibitory factor(s) in the AD brain. Mixing experiments, in which 50% of the control brain sample replaces an equal volume of an AD sample, show that the PN-1 activities are additive. Thus, the difference in PN-1 activity is not due to excess inhibitor (e.g., thrombin) in the AD sample or excess activator (e.g., heparan sulfate; Farrell and Cunningham 1986, 1987) in the control sample. The amount of hemoglobin in Alzheimer's and control samples (measured by CO difference spectroscopy) does not correlate with PN-1 levels (Wagner et al., 1989b). This indicates that the differences in PN-1 activity are not due to different levels of blood proteases that could form complexes with PN-1. Finally, the amounts of PN-1 protein estimated by the Western blots in the control brain samples approximate the amounts indicated by the above activity measurements. Thus, most of the PN-1 in the control samples is active.

If the blood-brain barrier is compromised in Alzheimer's disease as some reports suggest (Wisniewski and Kozlowski 1982; Hardy et al. 1986; Delacourte et al. 1987), this could lead to increased thrombin levels in the brain. Indeed, degenerating neurites have been observed around the periphery of angiopathic blood vessels in AD (Delacourte et al. 1987). The increased thrombin could form complexes with PN-1 leading to decreased levels of active PN-1. Measurements of thrombin in brain tissue, however, have not yet been carried out to test this possibility. Another possibility for decreased PN-1 in the AD brain would be decreased synthesis, though preliminary data suggest that PN-1 mRNA is not consistently reduced. Whatever the mechanism, the decline of PN-1 activity could lead to improper protease regulation which may contribute to AD pathology, e.g., neurite retraction.

Conclusions

It is now clear that the AD brain is capable of at least limited plasticity. The fact that sprouting can occur indicates that many of the biochemical mechanisms for restorative growth are operative. Nonetheless, in the AD brain it appears that regenerative growth can become intertwined in senile plaque formation. Local overproduction of neurotrophic factors and perhaps abnormal substrates may serve to stimulate aberrant local growth. Interestingly, the famous Spanish neuroanatomist Ramón y Cajal (1928) had already foreseen such mechanisms for plaque formation:

> *It appears as if the sprouts had been attracted towards the region of the plaque under the influence of some special neurotrophic substance.*

Thus, recent results suggest that, while many mechanisms regulating growth are intact in the AD brain, others, such as PN-1, are clearly abnormal. Reduced levels of PN-1 in AD together with increased thrombin levels could lead to disrupted interactions among neurites and altered neurite morphologies. Moreover, excessive protease action in the AD brain could also lead to abnormal fragments of other molecules (e.g., amyloid precursor protein) with their own aberrant biological (neurotrophic?) activity. Conversely, thrombin also inactivates a neurotrophic factor, acidic fibroblast growth factor (Lobb 1988), adding to potential complications. These possible mechanisms, compounded by others such as the susceptibility to excitotoxic injury, may make the AD brain particularly vulnerable to progressive injury.

In terms of transplantation into the AD brain, these and related considerations are relatively encouraging, yet also point toward extreme caution. Some of the possible competing mechanisms are summarized in Figure 2. The brain's environment will probably be able to support at least limited survival and growth. Indeed, even in highly degenerating brain regions normal neurons coexist with those inflicted with pathology. High protease activity along with the possibility that transplant fibers may become part of senile plaques, however, may limit the ultimate long-term therapeutic value of transplants. In AD, plasticity mechanisms, though initially adaptive, may participate in and even catalyze pathology. Thus, ultimately, plasticity comes at a very high cost.

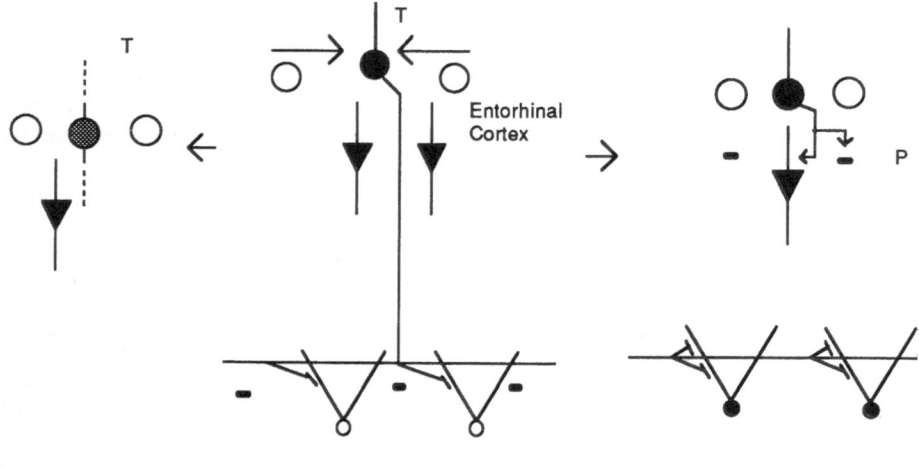

Dentate Gyrus

Fig. 2. Transplanted neurons may become involved in AD pathology. In order to replace cortical neurons into areas of degeneration, transplanted neurons must integrate into a complex network. This is indicated in the *center panel*. The transplanted neuron *(T)* must receive the proper input and project over a distance to its normal target neurons. It will fail to integrate completely if the fiber becomes part of plaques *(P)* or if the growing axon occupies only available synaptic sites on local neurons (see panel on *far right*). Alternatively (as shown on the *left panel*) a transplant neuron may degenerate due to a locally unfavorable environment. This could include excitotoxic insults due to a leaky blood-brain barrier, overactive proteases, or inadequate trophic influences

References

Abraham CR, Selkoe DJ, Potter H (1988) Immunochemical identification of the serine protease inhibitor α_1-antichymotrypsin in the brain amyloid deposits of Alzheimer's disease. Cell 52:487

Baker JB, Low DA, Simmer RL, Cunningham DD (1980) Protease-nexin: a cellular component that links thrombin and plasminogen activator and mediates their binding to cells. Cell 21:37—45

Baker JB, Knauer DJ, Cunningham DD (1986) Protease nexins: secreted protease inhibitors that regulate protease activity at or near the cell surface. In: Conn MP (ed) The receptors. Academic Press, New York, p 153

Ball MJ, Hachinski V, Fox A (1985) A new definition of Alzheimer's disease: a hippocampal dementia. Lancet I:14—16

Cotman CW, Nieto-Sampedro M (1984) Cell biology of synaptic plasticity. Science 255:1287—1294

Cotman CW, Matthews DA, Taylor D, Lynch G (1973) Synaptic rearrangement in the dentate gyrus: histochemical evidence for adjustments after lesions in immature and adult rat. Proc Natl Acad Sci USA 70:3473—3477

Delacourte A, Defossez A, Persuy P, Peers MC (1987) Observation of morphological relationships between angiopathic blood vessels and degenerative neurites in Alzheimer's disease. Virchows Arch [A] 411:199—204

Farrell DH, Cunningham DD (1986) Human fibroblasts accelerate the inhibition of thrombin by protease nexin. Proc Natl Acad Sci USA 83:6858—6862

Farrell DH, Cunningham DD (1987) Glycosaminoglycans on fibroblasts accelerate thrombin inhibition by protease nexin-1. Biochemistry 245:543—550

Farrell DH, Wagner SL, Yuan RH, Cunningham DD (1988) Localization of protease nexin-1 on the fibroblast extracellular matrix. J Cell Physiol 134:179

Geddes JW, Monaghan DT, Cotman CW, Lott IT, Kim RC, Chui HC (1985) Plasticity of hippocampal circuitry in Alzheimer's disease. Science 230:1179—1181

Geddes JW, Anderson KJ, Cotman CW (1986) Senile plaques as aberrant sprout stimulating structures. Exp Neurol 94:767—776

Gibbs RB, Harris EW, Cotman CW (1985) Replacement of damaged cortical projections by homotypic transplants of entorhinal cortex. J Comp Neurol 237:47—64

Gloor S, Odink K, Guenther J, Hanspeter N, Monard D (1986) A glia derived neurite promoting factor with protease inhibitory activity belongs to the protease nexin. Cell 47:687—693

Gurwitz D, Cunningham DD (1988) Thrombin modulates and reverses neuroblastoma neurite outgrowth. Proc Natl Acad Sci USA 85:3440—3444

Hardy JA, Mann DMA, Wester P, Winblad B (1986) An integrative hypothesis concerning the pathogenesis and progression of Alzheimer's disease. Neurobiology 7:489—502

Hoff SF, Scheff SW, Benardo LS, Cotman CW (1982) Lesion-induced synaptogenesis in the dentate gyrus of aged rats. I. Loss and reacquisition of normal synaptic density. J Comp Neurol 205:246—252

Hyman BT, van Hoesen GW, Damasio AR, Barnes CL (1984) Alzheimer's disease: cell-specific pathology isolates the hippocampal formation. Science 225:1168—1170

Hyman BT, van Hoesen GW, Kromer LJ, Damasio AR (1986) Perforant pathway changes and the memory impairment of Alzheimer's disease. Ann Neurol 20:472—481

Hyman BT, Kromer LJ, van Hoesen GW (1987) Reinnervation of the hippocampal perforant pathway zone in Alzheimer's disease. Ann Neurol 21:259—267

Kemper TL (1978) Senile dementia: a focal disease in the temporal lobe. In: Nandy K (ed) Senile dementia: A biomedical approach. Elsevier/North-Holland, Amsterdam, pp 105—113

Kitaguchi N, Takahashi Y, Tokushima Y, Shiojiri S, Ito H (1988) Novel precursor of Alzheimer's disease amyloid protein shows protease inhibitory activity. Nature 331:530—532

Lee KS, Stanford EJ, Cotman CW, Lynch GS (1977) Ultrastructural evidence for bouton proliferation in the partially deafferented dentate gyrus of the adult rat. Exp Brain Res 29:475—485

Lobb RR (1988) Thrombin inactivates acidic fibroblast growth factor but not basic fibroblast growth factor. Biochemistry 27:2572—2578

Low DA, Baker JB, Koonce WC, Cunningham DD (1981) Released protease-nexin regulates cellular binding, internalization, and degradation of serine proteases. Proc Natl Acad Sci USA 78:2340

Lynch GS, Matthews DA, Mosko S, Parks T, Cotman CW (1972) Induced acetylocholinesterase-rich layer in rat dentate gyrus following entorhinal lesions. Brain Res 42:311–318

Matthews DA, Cotman CW, Lynch G (1976a) An electron microscopic study of lesion-induced synaptogenesis in the dentate gyrus of the adult rat. I. Magnitude and time course of degeneration. Brain Res 115:1–21

Matthews DA, Cotman CW, Lynch G (1976b) An electron microscopic study of lesion-induced synaptogenesis in the dentate gyrus of the adult rat. II. Reappearance of morphologically normal contact. Brain Res 115:23–41

Monard D, Niday E, Limat A, Solomonson F (1983) Inhibition of protease activity can lead to neurite extension in neuroblastoma cells. Prog Brain Res 58:359–364

Nadler JV, Cotman CW, Lynch GS (1973) Altered distribution of choline acetyltransferase and acetylcholinesterase activities in the developing rat dentate gyrus following entorhinal lesions. Brain Res 63:215–230

Ponte P, Gonzalez-De Whitt P, Schilling J, Miller J, Hsu D, Greenberg B, Davis K, et al. (1988) A new A4 amyloid mRNA contains a domain homologous to serine proteinase inhibitors. Nature 331:525–527

Ramón y Cajal S (1928) Degeneration and regeneration of the nervous system. Hafner, New York

Rosenblatt DE, Cotman CW, Nieto-Sampedro M, Rowe JW, Knauer DJ (1987) Identification of a protease inhibitor produced by astrocytes that is structurally and functionally homologous to human protease nexin-1. Brain Res 415:40–48

Scott RW, Bergman BL, Bajpai A, Hersh RT, Rodriguez H, Jones BN, Barreda C, et al. (1985) Protease nexin: properties and a modified purification procedure. J Biol Chem 260:7029

Seress L, Ribak CE (1984) Commissural axons synapse with identified basket cells in rat dentate gyrus: a combined degeneration, Golgi-electron microscopic study. J Neurocytol 13:215–225

Tanzi RE, McClatchey AI, Lamperti ED, Villa-Komaroff L, Gusella JF, Neve RL (1988) Protease inhibitor domain encoded by amyloid protein precursor mRNA associated with Alzheimer's disease. Nature 331:528–530

Van Nostrand WE, Cunningham DD (1987) Purification of protease nexin II from human fibroblasts. J Biol Chem 262:8508–8514

Wagner SL, Geddes JW, Cotman CW, Lau AL, Isackson PJ, Cunningham DD (1988) Protease nexin-1, an antithrombin with neurite outgrowth activity, is reduced in Alzheimer's disease. #500.12. Soc. Neurosci Abstr 14:1247

Wagner SL, Geddes JW, Cotman CW, Lau AL, Isackson PJ and Cunningham DD (1989b) Protease nexin-1, an antithrombin with neurite outgrowth activity, is reduced in Alzheimer's disease (submitted)

Wagner SL, Lau AL, Cunningham DD (1989a) Binding of protease nexin-1 to the fibroblast surface alters its target proteinase specificity. J Biol Chem 264:611–615

Wisniewski HM, Kozlowski PB (1982) Evidence for blood brain barrier changes in senile dementia of the Alzheimer type (SDAT). Ann NY Acad Sci 396:119–129

Zurn AD, Nick H, Monard D (1988) A glia-derived nexin promotes neurite outgrowth in cultured chick sympathetic neurons. Dev Neurosci 10:17–24

Host Immune Responses to Neural Transplantation: A Possible Model for Autoimmune Diseases of the Central Nervous System

R. D. Lund, K. Rao, and *T. J. Gill*

Summary

Grafts of embryonic mouse neural tissue can survive transplantation to neonatal rat brains. Some grafts, however, undergo spontaneous rejection by the host immune system, showing that immune privilege in the brain is not absolute. Rejection can also be induced by a peripheral skin graft of mouse origin or by neural degeneration in the region of the graft. In each case, MHC antigen expression is induced in nonneuronal cells, and it is thought that this is responsible for initiating graft rejection. It is suggested that certain degenerative diseases of the brain may follow a similar pattern. Foreign antigens may be inserted into neural cell surfaces as a result of an infection or other transformation. Like the grafts, such cells may survive and function normally until confronted with a set of circumstances in which major histocompatibility complex (MHC) expression is induced. As a result, a sequence of degenerative events may then ensue.

Introduction

There are a number of degenerative disorders of the central nervous system (CNS) which share certain common features. The initial events that provoke the degenerative phase of the diseases are unknown, and as a result, the temporal relationship between the provocative event and the manifestation of the clinical syndrome is unknown. However, on the basis of such correlations as that between the influenza epidemic of 1918–1919 and an increased incidence of Parkinson's syndrome much later (Adams and Victor 1985), it is possible in some cases at least that the disease may not manifest itself until long after the event that induces it. The progress of the diseases is often protracted, quite erratic, and may involve periods of remission. The diseases are often recognized by a deficit in one system. Thus, most attention in Parkinson's syndrome has been directed to the motor deficits associated with the loss of the dopamine cells of the substantia nigra; in Alzheimer's disease, on the other hand, the behavioral deficits associated with the loss of the cholinergic cells in the basal forebrain have become a major focus of interest (Davies and Maloney 1976). In both cases, however, it is clear that other transmitter systems are also affected. In Parkinson's syndrome, norepinephrine and serotonin systems may also be affected, while an even broader array of transmitter systems are disturbed in Alzheimer's disease (Hamill et al. 1988). A major difficulty is in deciding which of the variety of neuropathological events is primary and which are secondary, since at the time the brain is examined, the disease is usually at an advanced stage. This makes it extremely difficult to define the causative agent and, for most of the diseases, a wide variety of factors have been implicated, including genetic, environmental, and viral. The possibility has often been suggested that an immune response may play a role in the course of the disease. Again the issue of whether such a response might be the cause or the result of the degenerative events is unknown.

One of the difficulties in approaching the problem of suspected autoimmune disorders in the central nervous system is providing the appropriate experimental models. Experimental allergic encephalomyelitis has been examined extensively as one such model (Paterson 1971). We propose here that transplantation of neural tissue to the brain may provide a particularly valuable model in which to examine and to manipulate the host immune system's response to foreign antigens introduced into the brain. The conditions under which the transplants undergo rejection may be analogous to those in which neural degeneration is provoked in certain disease states, and by studying them, it may be hoped that new insight will be provided as to the course and management of degenerative disease of the CNS.

An important factor in immune responses in the brain is the concept of immune privilege as it applies to the central nervous system. Even before the mechanisms of immune responsiveness had been defined, it was found that allografts of mouse carcinomas were rejected when placed in subcutaneous locations, but that they survived when transplanted into the brain of adult hosts (Ebeling 1914). In 1923, Murphy and Sturm reported the rapid growth of a mouse tumor graft that had been transplanted to a rat brain and concluded that this growth was due to the absence of a cellular reaction around the graft. They proposed that even though lymphoid cells surrounded blood vessels in the vicinity of the graft, the environment of the brain was not conducive to the migration of these cells into the parenchyma. Several similar studies culminated in the work of Medawar (1948) who suggested that the brain was a privileged site and that this privileged status had an immunological basis.

The concept of immune privilege has sometimes been interpreted as a total protection for immunogenetically mismatched tissue placed in the brain, but work done over the past 40 years has shown that while grafts can survive in the brain for prolonged periods, survival is not guaranteed (Medawar 1948; Scheinberg et al. 1966; Raju and Grogan 1977; Barker and Billingham 1977; Geyer and Gill 1979). Furthermore, the rejection process can be quite protracted compared with the response to similar tissue placed outside the central nervous system.

A variety of postulates has been proposed for the reduced responsiveness of the immune system to foreign tissue placed in the brain. They center around the relative isolation of the brain from the cells of the immune system and the associated relatively ineffective afferent limb of the immune response in the central nervous system (Barker and Billingham 1977). Thus, lymphatic drainage of the brain is extremely limited. However, areas where lymphatic drainage is present such as around the olfactory bulb are not noted as special targets for immune destruction. A second feature that has attracted attention is the fact that the blood-brain barrier prevents free diffusion of a variety of molecules from the vascular system into the brain parenchyma (Bradbury 1979) and may serve to limit the movement of lymphocytes into the brain. Again, there are a number of sites where the barrier is defective, yet these areas do not seem especially susceptible to immunological problems.

It has recently been documented that antigens must be presented to T lymphocytes as small peptides and that this presentation often involves cell surface molecules encoded by the major histocompatibility complex (MHC) (Watts and McConnell 1986; Unanue and Allen 1987). While MHC antigens are present on cells in blood vessel walls running through the brain, they are normally absent from both the neurons and neuroglia within the brain (Edidin 1977; Williams et al. 1980; Hart and Fabre 1981). This is likely to limit recognition of foreign antigens in the brain. Recent studies have indicated that activated T cells can cross the blood-brain barrier freely (Wekerle et al. 1986), but in general, immuno-histochemical analyses have shown very few T cells wihtin the brain of normal, healthy animals or humans. Thus, the lack of cells expressing MHC antigens together with a paucity of circulating T cells in brain parenchyma may be the central components of the immune protection afforded the brain.

There are, however, instances in which immune protection is not absolute. Several studies involving transplantation of nonneural tissue to the brain have shown that graft rejection occurs if donor and host MHC antigens differ (Geyer and Gill 1979; Head and Griffin 1985). The recent literature involving transplantation of neural tissue adds further support to the finding that grafts are susceptible to immune attack (Mason et al. 1986; Lund et al. 1987, 1988; Nicholas et al. 1987). These observations are of special importance when investigating the underlying mechanisms. If, for example, lack of MHC antigen expression in the brain affords immune protection to the brain cells, then is expression induced under conditions in which rejection occurs? There are a number of reports of brain cells expressing MHC antigens in a variety of degenerative diseases (Traugott et al. 1985; Lampson 1987; Sobel et al. 1987; Olsson et al. 1987), although it is not clear what causal relationship this plays, if any, to the degenerative process and what stimulates MHC expression under these circumstances. Experimental studies have shown that astrocytes in vitro and in vivo can be induced to express MHC antigens by administration of interferon-γ (Wong et al. 1984; Tedeschi et al. 1986), and it is possible that other diffusible factors may have similar effects. Once these cells express MHC antigens they can function as antigen-presenting cells and can present foreign antigens that may have entered the nervous system on some prior occasion, perhaps as the result of an earlier viral infection, and remained there unrecognized. Therefore, the two key issues in using transplantation as a model system for immune responses in the brain are the delineation of the conditions under which transplants undergo rejection and determination of whether MHC antigen expression plays a role in this process.

Results

We examine here the conditions under which grafts, either allografts between different strains of rat or xenografts between mouse and rat, are rejected and relate these to the cellular events that accompany the rejection process.

Xenografts — Neonatal Hosts

Tissue has been transplanted in a number of different studies from embryonic mouse central nervous system to the brains of neonatal rats. The cells in the grafts differentiate on a relatively normal timetable to provide a structure showing many of the characteristics of the donor region at maturity. The anatomically proper connections are made between graft and host brain and the graft can modify host neural functions in an appropriate manner (Hankin and Lund 1987; Klassen and Lund 1988). While these studies have suggested good graft survival, there are few systemic studies following graft survival with time. We have found that although embryonic mouse retinae can survive for as long as 14 months in a host rat brain, there is always a percentage of grafts at any survival time infiltrated with lymphocytes and a smaller group in which there is evidence of advanced rejection (Lund et al. 1988). The incidence of infiltration or overt rejection does not appear to increase systematically with age and it often varies dramatically among litters of rats of the same age, suggesting perhaps that certain extrinsic factors may play a role in precipitating an immune response. If all the grafts infiltrated with lymphocytes underwent rejection, it would be expected that with increased post-transplantation survival, there would be a higher proportion of brains without grafts. This, however, was not the case — opening the possibility that some infiltrated grafts may recover.

Embryonic cerebral cortex can show a similar pattern of survival. In recent studies, however, we have found that up to 4 weeks after placing a graft in a neonatal rat brain, less than 15% of grafts show evidence of rejection, but then at this age there is a sudden increase in rejection rate up to 85% of the grafts (Marion and Lund, unpublished result). It is not clear whether this dramatic change is due to maturation of a factor that modulates the immune response, to some change in the graft, or to another event, as yet to be defined.

Xenografts — Adult Hosts

We have found that if the recipient rats are more than about 8 days old at the time of transplantation, acute rejection is the usual response (Lund et al. 1987). Other studies grafting mouse tissue to the brains of adult rats show survival rates varying between near 0% (Inoue et al. 1985) and 56% (Björklund et al. 1982). It is not at all clear why there should be so much difference among the various studies, but there are many possible factors, including graft origin, graft type (solid or dissociated), graft location, amount of damage incurred during the process of introduction of the graft, and the strains of donor and recipient animals used. One point made in several studies is that survival can be improved by using the immunosuppressant cyclosporin A (Brundin et al. 1985, Inoue et al. 1985).

Allografts

Embryonic rat neural tissue frequently survives well after transplantation to the brains of neonatal rats. However, if host and donor have major immunogenetic disparities, a proportion of grafts undergo spontaneous rejection (Rao et al. 1987).

Allografts to adult recipients using defined donor and recipient strains have variable survival depending upon MHC and, to a lesser extent, non-MHC differences. If donor and host differed only at the MHC locus, survival for at least 60 days was common. Differences only at non-MHC loci also resulted in prolonged survival although lymphocytic infiltration was noted. If the donor and host varied in both MHC and non-MHC loci then rejection was much more common (Mason et al. 1986). A study in mice showed that class II MHC incompatibility alone led to rejection (Nicholas et al. 1987).

An important observation in the rat allograft studies (as well as in associated xenograft experiments) was the expression of class I MHC antigens on rejecting grafts (Mason et al. 1986). Nicholas et al. (1987) reported class II expression in and around rejecting grafts, although most of the cells showing the reaction seemed not to be endogenous to the CNS; Mason et al. (1986) were unable to demonstrate class II expression in cells intrinsic to their grafts. Many of the other studies on graft survival in adults used immunogenetically undefined strains, and, while of considerable value in addressing neurobiological questions, they provide little insight into the immunological mechanisms.

Induced Rejection

Skin Grafts

While neural allografts and xenografts placed in neonatal or adult rats and xenografts placed in the brains of neonatal rats may survive for prolonged periods, rejection of these grafts occurs if a skin graft from the donor strain is placed on the flank of the recipient animal (Freed 1983; Mason et al. 1986; Nicholas et al. 1987; Lund et al. 1987, 1989a, b). A series of cellular events accompanies this process (Houston et al. 1987; Lund et al. 1989b). The first event to be detected is an increase in the number of glial fibrillary acidic protein (GFAP)-positive astrocytes in and around the graft. Using species-specific antibodies, it was found that xenografts of cerebral cortex origin contain astrocytes of both host and donor origin. Early in the rejection process, however, the astrocytes of donor origin can no longer be detected. In an intermediate period, the remaining astrocytes hypertrophy, but whether they survive graft destruction is not yet clear. Five days after skin grafting, class II MHC antigens are heavily expressed on some cells in the graft, thought to be astrocytes, and the level of expression decreases over the succeeding days. Class I MHC antigen expression is first detected on day 6 and increases on succeeding days. Since antigen expression is most prominent at times when there is massive invasion of lymphocytes, it is difficult at the light microscopic level to be certain which cells are involved.

At 5 days after skin grafting, the first signs of leakage through the blood brain barrier are detected (Young et al. 1988). By 6 days, there is massive leakage both in and around the graft, and large numbers of lymphocytes and macrophages progressively invade the grafts until by 12—13 days the graft is largely destroyed.

After transplanting neural tissue from two species to a host brain it was found that a skin graft derived from one of the species caused the rejection of the neural graft derived from the same species only (Sefton and Lund 1988; Lund et al. 1988). This finding indicates that there is specificity in the rejection response.

Using congenic strains of rat, we were able to show that neural graft rejection depended on both MHC and non-MHC differences. Furthermore, our results indicated that the onset and progress of the rejection process were slower if the skin and the neural grafts varied only at non-MHC loci (Rao et al. 1989a, b). Skin grafts from third-party rat strains had no effect on the survival pattern of neural grafts.

Neural Lesions

Approximately 70% of mouse retinal grafts placed in the midbrain of neonatal rats survive for up to 90 days. If at 1 month or more of age, one or both eyes are removed from the host, 89% of those grafts lying in the dorsal part of the midbrain show signs of rejection (Lund et al. 1988). The rejection rate is in the normal range for more ventrally placed grafts. It would appear, therefore, that degeneration of optic terminals in the dorsal midbrain, caused by the eye removal precipitates a set of events that leads to rejection of a graft in the vicinity. In order to determine the stimulus for this rejection, we have removed eyes from normal adult rats without grafts to see what changes accompany the degeneration of the optic terminals. There is clearly an increase in GFAP-positive astrocytes as well as hypertrophy of microglia demonstrable with OX42 and W3/25 antibodies. The most surprising result, however, is that between 2 and 3 days after eye removal a group of cells (some, at least, of which show certain characteristics of astrocytes) stain with antibodies for class I (OX18) and class II (OX6) MHC antigens (Rao and Lund 1988). This staining persists for at least 3 weeks. How this impacts on the survival of grafts in the region remains to be examined.

Discussion

The cells in the neural grafts described here, by virtue of the connections they make and the functions they mediate, express important characteristics of the cells in the regions from which the grafts were derived. These characteristics presumably include cell surface molecules thought to be important in guiding neurite outgrowth and various other cellular interactions including synapse formation. The membranes do, however, contain certain species-specific antigens, thereby distinguishing them from membranes of cells in the host brain (Charlton et al. 1983; Lund et al. 1985, 1989b). Despite this, they can survive and function in the host brain for prolonged periods, presumably because in the absence of MHC expression in cells in the brain, the foreign antigens cannot be presented to T lymphocytes.

In some ways allograft and xenograft cells may be compared to cells in a normal brain whose surface membranes contain new molecules resulting from viral infections or other transformations. These too may go unrecognized in the absence of appropriate antigen-presenting cells. However, if conditions prevail which induce MHC expression on cells in the brain, then rejection may occur. It is clear that graft rejection can sometimes be a very slow process. Furthermore, statistics on graft survival at different ages taken together with the percentage of grafts that are infiltrated with lymphocytes suggest that some grafts that are infiltrated at an early age may subsequently recover. Whether they deteriorate during the period of infiltration is unknown, but it is our experience that retinal xenografts generally function less well with age than do retinal allografts. One point of significance is that some astrocytes appear to suppress T-helper cell and spleen mononuclear cell proliferation in vitro (Borgeson and Keane 1988); if this were to happen in vivo, it could represent a potent mechanism for regulating or limiting the rejection process. The observations and extrapolations presented here may be relevant in explaining the progress of certain human degenerative diseases of the central nervous system, where 1. a transmissible agent may be involved; 2. MHC expression is found on cells in the brain parenchyma; and 3. evidence exists for the participation of immunoresponsive mechanisms, including overt changes such as lymphocytic invasion and breakdown of the blood-brain barrier.

These conditions are clearly met in a number of diseases including a human disease closely associated with Alzheimer's disease, Creutzfeldt-Jakob disease, and a commonly used animal model for Alzheimer's disease, scrapie infected mice. However, the situation in Alzheimer's disease is somewhat less clear although circumstantial and more direct evidence is beginning to point to a similar etiology. An association between Alzheimer's disease and a transmissible agent such as a virus or prion has frequently been suspected (Brown et al. 1982; Salazar et al. 1983; Carp et al. 1984; Wisniewski et al. 1985), although a number of attempts to demonstrate such an agent have proven unsuccessful (Goudsmit et al. 1980; Manuelidis 1985). Recently, however, evidence for a transmissible agent from patients in early stages of development of the disease has been presented (Manuelidis et al. 1988). Several studies have shown an increased expression of class II MHC antigens, especially associated with microglia, in Alzheimer cases (McGeer et al. 1987; Pouplard-Barthelaix 1988; Mattiace et al. 1988). Finally, evidence is growing for an immune component to Alzheimer's disease. Nandy (1985), for example, correlates the pathological events occurring in aging with diminished immune function; Pouplard-Barthelaix (1988) proposed an autoimmune mechanism in provoking the degenerative event; Fudenberg and Singh (1988) suggest that a least one of the four subclasses of Alzheimer's disease is of immunological origin. Evidence for an incompetent blood-brain barrier has also been presented (Wisniewski et al. 1985; Pouplard-Barthelaix 1988).

It could, therefore, be proposed that an infective agent, perhaps a virus, is the primary cause of Alzheimer's disease, and that the disease is not manifested unless events similar to those leading to graft rejection occur in the brain. It is clear that a major factor is the role of the neuroglia in MHC-antigen expression and that future research both in degenerative diseases such as Alzheimer's disease and in transplant rejection lies in understanding how this process is modulated and regulated.

Acknowledgments. Supported by NIH grants EY05283 and CA18659, and by grants from the Samuel and Emma Winters Foundation and from the Emmerling Fund of the Pittsburgh Foundation

References

Adams RD, Victor M (1985) Principles of neurology. McGraw-Hill, New York, pp 560−563
Barker CF, Billingham RE (1977) Immunologically privileged sites. Adv Immunol 23:1−54
Björklund A, Stenevi U, Dunnett SB, Gage FH (1982) Cross-species neural grafting in a rat model of Parkinson's disease. Nature 298:652−654
Borgeson M, Keane RW (1988) Astrocyte-mediated immunosuppression. Soc Neurosci Abstr 14:757
Bradbury M (1979) The concept of a blood-brain barrier. Wiley, Chichester
Brown P, Salazar AM, Gibbs CJ, Gajdusek DC (1982) Alzheimer's disease and transmissible virus dementia (Creutzfeldt-Jakob disease). Ann NY Acad Sci 396:131−143
Brundin P, Nilsson OG, Gage FH, Björklund A (1985) Cyclosporin A increases survival of cross-species intrastriatal grafts of embryonic dopamine-containing neurons. Exp Brain Res 60:204−208
Carp RI, Merz GS, Wisniewski HM (1984) Transmission of unconventional slow virus diseases and the relevance to AD/SDT transmission studies. In: Wertheimer J, Marois M (eds) Senile dementia: outlook for the future. Liss, New York, pp 31−54
Charlton HM, Barclay AN, Williams AF (1983) Detection of neuronal tissue from brain grafts with anti-Thy-1.1 antibody. Nature 305:825−827
Davies P, Maloney AJR (1976) Selective loss of central cholinergic neurones in Alzheimer's disease. Lancet II:1403
Ebeling E (1914) Experimentelle Gehirntumoren bei Mäusen. Z Krebsforsch 14:151−156
Edidin M (1972) The tissue distribution and cellular location of transplantation antigens. In: Kahan BD, Reisfeld R (eds) Transplantation antigens. Academic, New York, pp 125−140
Freed WJ (1983) Functional brain tissue transplantation: reversal of lesion-induced rotation by intraventricular substantia nigra and adrenal medulla grafts, with a note on intracranial retinal grafts. Biol Psychiatry 18:1205−1267
Fudenberg HH, Singh VK (1988) Immunodiagnosis and immunotherapy of patients with Alzheimer's syndrome. In: Pouplard-Barthelaix A, Emile J, Christen Y (eds) Immunology and Alzheimer's disease. Springer, Berlin Heidelberg New York, pp 98−107
Geyer SJ, Gill TJ III (1979) Immunogenetic aspects of intracerebral skin transplantation in inbred rats. Am J Pathol 94:569−584
Goudsmit J, Morrow CH, Asher DM, Yanagihara RT, Masters CL, Gibbs CJ, Gajdusek DC (1980) Evidence for and against the transmissibility of Alzheimer disease. Neurology (NY) 30:945−950
Hamill RW, Caine E, Eskin T, Lapham L, Shoulson I, McNeill TH (1988) Neurodegenerative disorders and aging. Alzheimer's disease and Parkinson's disease − common ground. Ann NY Acad Sci 515:411−419
Hankin MH, Lund RD (1987) Role of the target in directing the outgrowth of retinal axons: transplants reveal surface-related and surface-indepedent cues. J Comp Neurol 263:455−466

Hart DNJ, Fabre JW (1981) Demonstration and characterization of Ia-positive dendritic cells in the interstitial connective tissue of rat heart and other tissues but not brain. J Exp Med 153:347−361

Head JR, Griffin WST (1985) Functional capacity of solid tissue transplants in the brain: evidence for immunological privilege. Proc R Soc Lond [Biol] 224:375−387

Houston MB, Kunz HW, Gill TJ III, Lund RD (1987) Course of transplant rejection within the CNS. Soc Neurosci Abstr 13:288

Inoue H, Kohsaka S, Yoshida K, Ohtani M, Toya S, Tsukada Y (1985) Cyclosporine A enhances the survivability of mouse cerebral cortex grafted into the third ventricle of the rat brain. Neurosci Lett 54:85

Klassen H, Lund R (1988) Anatomical and behavioral correlates of a xenograft-mediated pupillary reflex. Exp Neurol 102:102−108

Lampson LA (1987) Molecular bases of the immune response to neural antigens. Trends in Neurosci 10:211−216

Lund RD, Chang F-LF, Hankin MH, Lagenaur CF (1985) Use of a species-specific antibody for demonstrating mouse neurons transplanted to rat brains. Neurosci Lett 61:221−226

Lund RD, Rao K, Hankin MH, Kunz HW, Gill TJ III (1987) Transplantation of retina and visual cortex to rat brains of different ages: maturation, connection patterns and immunological consequences. Ann NY Acad Sci 495:227−241

Lund RD, Rao K, Kunz HW, Gill TJ III (1988) Instability of neural xenografts placed in neonatal rat brains. Transplantation 46:216−223

Lund RD, Rao K, Kunz HW, Gill TJ III (1989a) Immunological considerations in neural transplantation. Transplant Proc (in press)

Lund RD, Houston MB, Lagenaur CF, Kunz HW, Gill TJ III (1989b) Cellular events associated with induced rejection of neural xenografts placed in neonatal rat brains. Transplant Proc (in press)

Manuelidis EE (1985) Creutzfeldt-Jakob disease. J Neuropathol Exp Neurol 44:1−17

Manuelidis EE, de Figueiredo JM, Kim JH, Fritch WW, Manuelidis L (1988) Transmission studies from blood of Alzheimer disease patients and healthy relatives. Proc Natl Acad Sci USA 85:4898−4901

Mason DW, Charton HM, Jones AJ, Lavy CBD, Puklavec M, Simmonds SJ (1986) The fate of allogeneic and xenogeneic neuronal tissue transplanted into the third ventricle of rodents. Neuroscience 19:685−694

Mattiace LA, Gong G, Dickson DW (1988) HLA-DR-positive microglia in normal and diseased brains. Soc Neurosci Abstr 14:1086

McGeer PL, Itagaki S, Tago H, McGear EG (1987) Reactive microglia in patients with senile dementia of the Alzheimer type are positive for the histocompatibility glycoprotein HLA-DR. Neurosci Lett 79:195−200

Medawar PB (1948) Immunity to homologous grafted skin. II. The fate of skin homografts transplanted to the brain, to subcutaneous tissue, and to the anterior chamber of the eye. Br J Exp Pathol 29:58−69

Murphy JE, Sturm E (1923) Conditions determining the transplantability of tissues in the brain. J Exp Med 38:183−197

Nandy K (1985) Immunopathology of aging and dementia. In: Hutton JT, Kenny AD (eds) Senile dementia of the Alzheimer type. Liss, New York, pp 293−305

Nicholas MK, Antel JP, Stefansson K, Arnason BGW (1987) Rejection of fetal neocortical neural transplants by H-2 incompatible mice. J Immunol 139:2275−2283

Olsson T, Maehlen J, Löve A, Klareskogl L, Norrby E, Kristensson K (1987) Induction of class I and class II transplantation antigens in rat brain during fatal and non-fatal measles virus infection. J Neuroimmunol 16:215−224

Paterson PY (1971) Experimental allergic encephalomyelitis and autoimmune disease. Adv Immunol 5:131−140

Pouplard-Barthelaix A (1988) Immunological markers and neuropathological lesions in Alzheimer's disease. In: Pouplard-Barthelaix A, Emile J, Christen Y (eds) Immunology and Alzheimer's disease. Springer, Berlin Heidelberg New York, pp 7−16

Raju S, Grogan JP (1977) Immunologic study of the brain as a privileged site. Transplant Proc 9:1187−1191

Rao K, Lund RD (1988) Cellular response to optic terminal degeneration. Soc Neurosci Abstr 14:583

Rao K, Kunz HW, Gill TJ III (1987) Involvement of MHC antigens in neural allograft rejection. Soc Neurosci Abstr 13:288

Rao K, Kunz HW, Gill TJ III, Lund RD (1989a) MHC-dependent neural allograft rejection. Ann NY Acad Sci 540:493−494

Rao K, Lund RD, Kunz HW, Gill TJ III (1989b) Immunological implications of xenogeneic and allogeneic transplantation to neonatal rats. Prog Brain Res 78:281−286

Salazar AM, Brown P, Gajdusek DC, Gibbs CJ (1983) Relation to Creutzfeldt-Jakob disease and other unconventional virus diseases. In: Reisberg B (ed) Alzheimer's disease. Macmillan, London, pp 311−318

Scheinberg LC, Kotsilimbas DG, Karpf R, Mayer N (1966) Is the brain "an immunologically privileged site"? III. Studies based on homologous skin grafts to the brain and subcutaneous tissues. Arch Neurol 15:62−67

Sefton AJ, Lund RD (1988) Co-transplantation of embryonic mouse retina with tectum, diencephalon or cortex to neonatal rat cortex. J Comp Neurol 269:548−564

Sobel RA, Natale JM, Schneeberger EE (1987) The immunopathology of acute experimental allergic encephalomyelitis. IV. An ultrastructural immunocytochemical study of class II major histocompatibility complex molecule (Ia) expression. J Immunopathol Exp Neurol 46:239−249

Tedeschi B, Barrett JN, Keane RW (1986) Astrocytes produce interferon that enhances the expression of H-2 antigens on a subpopulation of brain cells. J Cell Biol 102:2244−2253

Traugott V, Scheinberg LC, Raine CS (1985) On the presence of Ia-positive endothelial cells and astrocytes in multiple sclerosis lesions and its relevance in antigen presentation. J Neuroimmunol 8:1−14

Unanue ER, Allen PM (1987) The basis for the immunoregulatory role of macrophages and other accessory cells. Science 236:551−557

Watts TH, McConnell HM (1986) High-affinity fluorescent peptide binding to Ia^d in lipid membranes. Proc Natl Acad Sci USA 83:9660−9664

Wekerle H, Livingston C, Lasman H, Meyermann R (1986) Cellular immune reactivity within the CNS. Trends Neurosci 9:271−277

Williams K, Hart D, Fabre J, Morris P (1980) Distribution and quantitation of HLA-ABC and DR (Ia) antigens on human kidney and other tissues. Transplantation 29:274−279

Wisniewski HM, Merz GS, Wen GY, Iqbal K, Grundke-Iqbal I (1985) Morphology and biochemistry of Alzheimer's disease. In: Hutton JT, Kenny AD (eds) Senile dementia of the Alzheimer type. Liss, New York, pp 263−274

Wong GHW, Bartlett PF, Clark-Lewis I, Battye F, Schrader WJ (1984) Inducible expression of H-2 and Ia antigens on brain cells. Nature 310:688−691

Young MJ, Rao K, Lund RD (1989) Integrity of the blood-brain barrier in retinal xenografts is correlated with the immunological status of the host. J Comp Neurol (in press)

NGF in CNS: Sites of Synthesis and Effects of Novel Ways to Administer NGF on Intrinsic Cholinergic Neurons and Grafts of Cholinergic Neurons and Their Target Areas

L. Olson, C. Ayer-LeLièvre, T. Ebendal, M. Eriksdotter-Nilsson, P. Ernfors, P. Mouton, H. Persson, and *I. Strömberg*

Abstract

Recent evidence suggests that nerve growth factor (NGF) is present in the central nervous system and that cholinergic neurons are among those that respond to, and may depend on, NGF. We have recently been able to locate NGF mRNA to pyramidal and granular cells of the hippocampal formation and have proposed a neuron-to-neuron interaction between such NGF-producing cells and the cholinergic projections to the hippocampal formation. When fetal basal forebrain tissue rich in cholinergic neuroblasts is grafted to the anterior chamber of the eye, the grafts will reach a larger final volume and contain a larger number of both cholinergic and noncholinergic neurons if supported by repeated NGF injections into the eye chamber. Conversely, grafts of cerebral cortex and hippocampus will both stop growing earlier in the presence of exogenous NGF, suggesting that NGF inhibits cell division and initiates differentiation in these cholinergic target areas. Cholinergic cell-rich cell suspensions were also injected into the cerebral cortex of animals with ibotenic acid-induced lesions of nucleus basalis. Again, NGF, infused chronically using a dialysis fiber-osmotic minipump assembly, enhanced graft survival and cholinergic reinnervation of host cortex. A genetically engineered cell line that produces and secretes recombinant NGF has been tried as an alternative continuous source of NGF. When compared with the parent cell line, the NGF-oversecreting cells were found able to evoke cholinergic growth responses in intact cholinergic intrinsic neurons in both cortex and striatum, to rescue cholinergic neurons in the basal forebrain after fimbria-fornix lesions, and to stimulate grafts of fetal cholinergic neurons in the brain. Genetically engineered cells thus open up new possibilities for future treatment strategies in neuro-degenerative diseases, such as Alzheimer's disease, in which treatment with neurotrophic factors might prove beneficial.

Introduction

It was 82 years ago that Alois Alzheimer first described the classical neuropathology of the disease that bears his name (Alzheimer 1907). Recent evidence suggests that cholinergic neurons, in particular the projections from basal forebrain to cortex and from septum to hippocampus, may be important in memory, learning, and other cognitive functions and that these cholinergic projections undergo degenerative changes in aging (Bartus et al. 1982) and senile dementia of the Alzheimer type (Davies and Maloney 1976; Perry et al. 1977, 1982; White et al. 1977; Arendt et al. 1985; Bowen et al. 1979; Pearson et al. 1983; Whitehouse et al. 1982).

Interestingly, nerve growth factor (NGF) is present in the brain (see Whittemore et al. 1986), can stimulate cholinergic neurons (see Thoenen et al. 1987), and appears to decrease in concentration during aging (Lärkfors et al. 1987). In this chapter we shall review our recent studies of the localization of NGF mRNA in the brain and the possible role of NGF in the developing and adult central nervous system, as evidenced by grafting fetal brain tissue to the anterior chamber of the eye and to the brain of adult recipients and adding NGF. In addition, by transplanting genetically engineered cells that secrete substantial amounts of NGF, we will demonstrate effects of NGF on local cholinergic neurons in cortex and striatum and how such cells can be used to rescue injured cholinergic projection neurons in the CNS.

NGF mRNA Is Present in Hippocampal Neurons

Blot-hybridization has demonstrated an abundance of NGF mRNA in the hippocampus, cerebral cortex, and olfactory bulb, and enzyme immunoassays have revealed significant levels of NGF-like protein in the hippocampus and cortex cerebri of the adult brain (Whittemore et al. 1986). To carry these observations to the resolution of the cellular level, in situ hybridization with complementary DNA probes for NGF was used to identify cells that might contain NGF mRNA in the rat and mouse brain (Ayer-LeLièvre et al. 1988). Hybridization was carried out with a ^{32}P-labeled 900-bp Pst I fragment from a mouse NGF cDNA clone. This fragment shows 97% DNA sequence homology with the rat gene and hybridizes strongly to rat NGF mRNA under stringent conditions (Korsching et al. 1985; Whittemore et al. 1986; Shelton and Reichardt 1986). Strong labeling occurred in the dentate gyrus and in the hippocampal pyramidal layers in areas CA1 to CA4. To evaluate the specificity of this labeling, the probe was mixed with an excess of unlabeled pUC9 DNA to eliminate possible nonspecific labeling over the hippocampus. The same labeling pattern was seen with this mixture. Moreover, a nonoverlapping NGF cDNA fragment, covering the 3' untranslated region, showed an identical hybridization pattern, thus providing evidence against a nonspecific cross-hybridization of the probes with mRNA other than NGF mRNA. Further evidence for specificity of the labeling to hippocampal structures was provided by the absence of labeling with a 900 bp Pst I fragment from the chicken NGF gene, since there is very little cross-hybridization between mouse NGF mRNA and chicken NGF probes using RNA blot analysis or in situ hybridizations and high stringency conditions (Ayer-LeLièvre et al. 1989; Ebendal et al. 1986). Also, there was no hybridization when pUC9 DNA was used as a probe or after treatment of the sections with ribonuclease A. To prove that the in situ hybridization techniques as used by us gave "expected" patterns of labeling, the NGF probes were tested and found to generate very strong signals with the expected cellular distribution in the male mouse submandibular gland (Ayer-LeLièvre et al. 1989), whereas the reaction with a probe for the rat NGF receptor was negative in the hippocampal formation and distinctly positive in cells in the basal forebrain-septum areas such as the diagonal band of Broca.

Better cellular resolution was obtained with ^{3}H-labeled NGF cDNA probes. We could now demonstrate labeling in the cytoplasm of cell bodies with the typical location, shape and dimension of pyramidal neurons in the hippocampus as well as granular neurons in the gyrus dentatus. Computerized image analysis showed that the labeling over these neurons was 11- to 66-fold higher than background (Ayer-LeLièvre et al. 1988). Although there has been no detailed study with in situ hyridization outside the hippocampal formation, it is clear that cells in other areas such as the cortex cerebri also contain NGF mRNA (Ayer-LeLièvre et al. 1988).

We were unable to detect labeling with our NGF mRNA probe over areas rich in oligodendrocytes, such as corpus callosum. This, however, does not exclude levels of NGF mRNA below the detection limit of our method, or that NGF mRNA might increase in glial cells after lesions or other perturbations. When granular cells in the dentate gyrus were destroyed by local injections of colchicine, labeling with the NGF probe disappeared from the same areas. In parallel, as shown by RNA blot analysis of lesioned areas, there was a marked decrease in the neuronal marker Thy-1 mRNA, while the astrocytic marker glial fibrillary acidic protein mRNA increased, again strongly suggesting a neuronal rather than glial localization of the NGF message. Similarly, when hippocampal pyramidal neurons were selectively destroyed by local injections of kainic acid, the in situ hybridization with the NGF probe disappeared in precisely those areas of the hippocampal gyrus where pyramidal neurons were lost (Ayer-LeLièvre et al. 1988).

We conclude from these studies that NGF is produced in neurons in the brain such as the hippocampal pyramidal neurons and the granular cells of the dentate gyrus. These results support the notion that the cholinergic projections from septum to hippocampus are maintained by a continuous trophic supply of NGF from the target neurons through a neuron-to-neuron interaction. Disturbances of this interaction may underlie pathological alterations in aging and other diseases of the central nervous system.

NGF Stimulates Basal Forebrain Tissue Grafts in the Eye Chamber

By grafting defined areas of the developing central nervous system to the anterior chamber of the eye (see Olson et al. 1983), it becomes possible to evaluate the degree to which a given area can develop and function normally in complete isolation from the rest of the central nervous system and, to the extent that disturbances are found, how they might be counteracted by addition of other parts of the central nervous system in the form of sequential double or triple grafts or by addition of trophic factors.

We have recently used the anterior chamber grafting technique to further characterize the possible role of NGF in basal forebrain development (Eriksdotter-Nilsson et al. 1989a). Small pieces of fetal septum and diagonal band dissected from fetuses at embryonic day 17 were incubated in NGF or saline and grafted to the eye chamber. The eyes then received weekly injections of NGF (500 ng in 5 μl saline) or saline. Interestingly, NGF treatment caused an increased growth of the grafts evident 3 weeks postgrafting and remaining highly significant 6 weeks after grafting, when the volume of the NGF-treated grafts was approximately 60% larger than that of the saline-treated grafts. Both NGF and control grafts became rapidly vascularized from the host iris, and at histological examination after intraocular maturation they appeared to have similar densities of neurons and glial elements.

Three different markers for cholinergic neurons – acetylcholinesterase (AChE), histochemistry, and immunohistochemistry using antibodies against both AChE and choline acetyltransferase (ChAT) – were used to count the number of cholinergic neurons in the grafts and to estimate the density of cholinergic nerve fibers (Eriksdotter-Nilsson et al. 1989). The results clearly demonstrated an increased number of permanently surviving cholinergic neurons in the NGF-treated grafts. There was, however, no change in density of either cholinergic cell bodies or fibers caused by the NGF treatment. Thus, the number of cholinergic cells had increased in proportion to the volume increase of the whole graft, suggesting either that the stimulation of cholinergic survival had in turn led to stimulation of other components of the grafted parenchyma or, alternatively, that NGF also had direct effects on other, noncholinergic constituents of the grafts. Whatever the mechanism, these experiments show that NGF promotes survival of target-deprived cholinergic neurons and directly or indirectly also supports noncholinergic neurons during development.

NGF Can Influence Development of Cortex Cerebri and Hippocampus

As indicated above, NGF mRNA is found in neurons in the hippocampal formation and also in cells in cortex cerebri. While a major role for NGF synthesized in these regions may be to support the afferent cholinergic projections, it is conceivable that NGF might play a role locally in these cortical areas as well. To examine this possibility, we grafted fetal cortex cerebri from the parietal region and the developing hippocampal formation to the anterior chamber of the eye (Eriksdotter-Nilsson et al. 1989b). The grafts received NGF or a control solution containing cytochrome c (a commonly used control for NGF) at grafting and in the form of intraocular injections at day 5 and 10 postgrafting.

We found that NGF caused cortex cerebri grafts from both embryonic day 15 and day 17 to grow to significantly smaller final sizes than the cytochrome c-treated grafts. This effect of exogenous NGF was seen when 5-μl injections of NGF in a concentration of 100 μg/ml were used (Fig. 1a), and seemed to be dose dependent since small effects were still noted using 10 μg/ml. To control for the unlikely event that cytochrome c per se had any effects on graft growth, injections of this compound were compared with injections of saline in a separate experiment. There were no differences between these two control treatments.

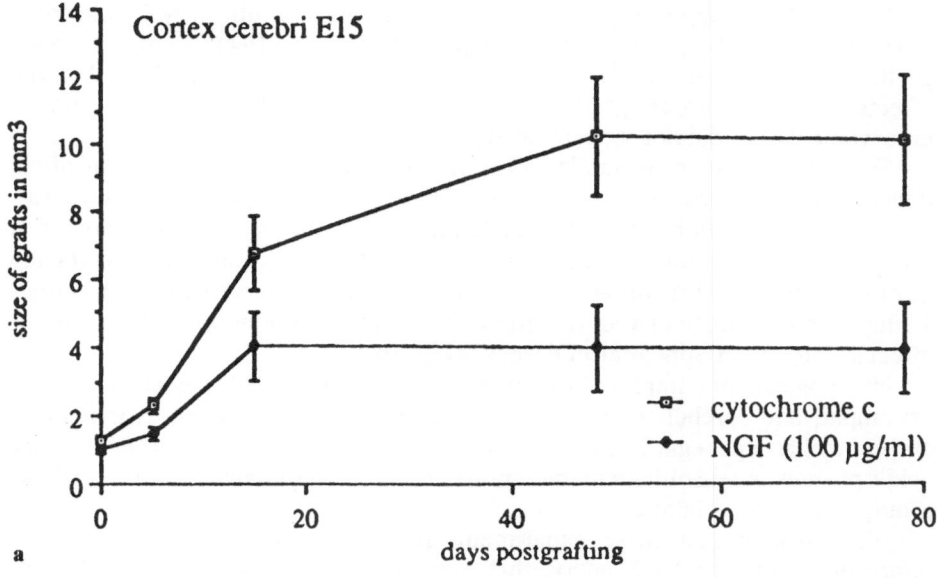

a

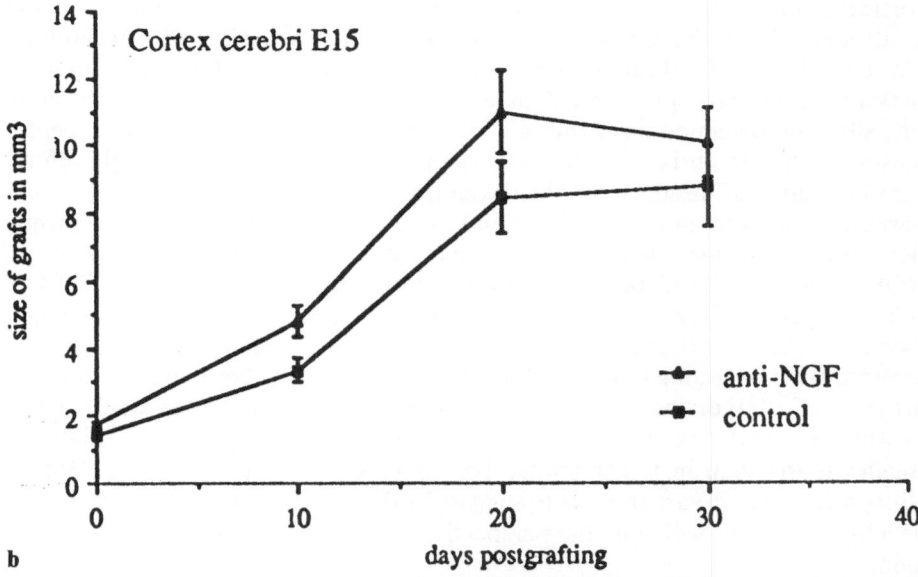

b

Fig. 1a, b. Growth curves of cortex cerebri grafts derived from gestational day 15. **a** The grafts were treated with NGF (100 µg/ml) or cytochrome c at grafting and days 5 and 10 post-transplantation. The cytochrome c-treated grafts ($n = 16$) grew significantly ($P < 0.01$, ANOVA) larger over time than the NGF-treated grafts ($n = 12$). This difference in growth is significant after 1 week in oculo. **b** The grafts received intraocular injections of anti-NGF every 5 days for 4 weeks, resulting during the first 3 weeks in a slightly larger size of the anti-NGF-treated cortex cerebri grafts (E15, $n = 14$) as compared with control grafts ($n = 16$) receiving injections of preimmune serum. After 4 weeks in oculo, this difference is no longer seen. (From Eriksdotter-Nilsson et al. 1989b)

Grafts of the hippocampal formation also grew to significantly smaller final sizes in oculo in the presence of exogenous NGF. The effects were seen using grafts from embryonic day 18 and were somewhat smaller in magnitude than the effects in cortex cerebri grafts. The growth of cerebellar grafts obtained at embryonic day 13 did not seem to be affected at all by NGF.

Several attempts were made to see if antibodies against NGF might affect development of cortical grafts. These included immunization of the host rats against NGF, injections of antibodies against NGF into the anterior chamber of the eye, and combinations thereof. In one case, that of cortical grafts receiving intraocular injections of antibodies, we found that the grafts were markedly larger 10 days postgrafting than a control group that received preimmune serum, but this effect seemed to disappear after 4 weeks (Fig. 1b).

Histological and immunohistochemical examinations of the cortical, hippocampal, and cerebellar grafts revealed no differences between experimental and control groups regarding general organization or distribution of glial fibrillary acidic protein or neurofilament immunoreactivity. Image analysis confirmed these semiquantitative estimations.

We conclude that these experiments have revealed a hitherto unknown interesting aspect of NGF action that may also be important in normal brain development. It seems unlikely that the early cessation of growth induced in the cortical grafts should be a toxic effect of NGF, since it is not seen in cerebellar grafts and since NGF has the opposite effect, a stimulation of growth, in cholinergic cell-rich basal forebrain grafts. A recent study has indeed demonstrated the presence of NGF receptor mRNA in hippocampus and cortex cerebri of the adult rat, albeit in low amounts (Ernfors et al. 1989b). The amount of NGF receptor message is higher during development (Ernfors et al. 1989b) and it is, therefore, possible that NGF produced in the developing hippocampus and cortex acts as a paracrine signal to end cell division and initiate differentiation, synapse formation, and maturation. Such an effect would not be unlike other known effects of NGF, such as inhibiting proliferation and inducing neuronal maturation in PC12 cells. Exogenous NGF in high amounts, as in the present set of experiments, might then have caused an exaggeration of this phenomenon, leading to precocious maturation. It is interesting to note that the cortical grafts that became smaller in the presence of NGF did not differ from the control grafts in degree of gliosis, while in other intraocular experiments the degree of gliosis is almost invariably higher in smaller grafts than in larger grafts. To our knowledge our data (Eriksdotter-Nilsson et al. 1989b) are the first to suggest NGF as a maturation signal in cortical development, thus widening the perspective of a role for NGF in brain development.

Chronic NGF Infusion Supports Basal Forebrain Grafts to the Brain

We have previously demonstrated that NGF can be administered chronically to the brain parenchyma using a subcutaneously implanted osmotic minipump attached by thin tubing to a stereotaxically implanted dialysis fiber (Strömberg et al. 1985), and that such NGF remains active in the pump for several weeks and can exert profound effects on grafts of chromaffin tissue (Strömberg et al. 1985). We have now used the same NGF infusion technique to stimulate grafts of cholinergic-rich cell suspensions prepared from embryonic day 17 basal forebrain and grafted to cortex cerebri of adult recipients. To be able to evaluate cholinergic fiber production in these experiments, we removed the cholinergic afferents to cortex cerebri of the host by ibotenic acid injections into nucleus basalis (Mouton et al., in preparation).

The cholinergic-rich cell suspensions aggregate and form solid grafts in cortex. If placed adjacent to a dialysis fiber delivering NGF, the grafts became markedly larger and contained more cholinergic neurons leading to an increased cholinergic reinnervation of host cortex, as compared with grafts being close to dialysis fibers delivering cytochrome c (Mouton et al., in preparation). Thus, the level of NGF present in cholinergically partially denervated (the intrinsic cholinergic neurons remain) adult cortex is not enough for maximal development of grafts of fetal cholinergic neurons. Experiments now underway are aimed at trying to define to what extent cholinergic grafts, supported by NGF, to the cholinergically denervated cortex might be able to reverse functional deficits caused by the denervation.

Grafts of Genetically Engineered Cells that Make NGF: Effects on Cholinergic Neurons in the Adult Brain

We have recently described the construction of a stable cell line, derived from mouse 3T3 cells, that produces and secretes large amounts of recombinant NGF (Olson et al. 1988; Ernfors et al. 1989a). A construct containing a DNA fragment from the rat NGF gene was transfected together with a plasmid-containing neomycin into mouse fibroblast 3T3 cells. Conditioned medium from cells resistant to the antibiotic G-418 was tested in an in vitro bioassay system for NGF activity, and cell lines were selected that secreted high amounts of biologically active NGF. One such cell line, 3E, contained several hundred copies of the rat NGF gene per cell and large amounts of NGF mRNA, and secreted NGF in the order of 5 ng per ml of medium as compared with 0.2 ng per ml produced by the parent 3T3 cell line (Ernfors et al. 1989a).

We have tested the ability of 3T3 and 3E cells to survive grafting to mice and to immunosuppressed rats. There were some initial problems with tumor formation with the somewhat heterogeneous parent 3T3 cell line. These problems were overcome by selection of subclones of the parent cell line. No tumor formation was noted with the 3E cells. Survival of the grafted cells was monitored by routine histology and by fibronectin immunohistochemistry. While cells survived when injected as cell suspensions, better survival seemed to occur when they were pre-cultured in a collagen gel which was then cut into pieces suitable for grafting.

When gels carrying NGF-secreting 3E cells were grafted into the center of striatum of cyclosporin-treated rats, they caused a pronounced local cholinergic hyperinnervation of adjacent host striatal neuropil within a few weeks (Fig. 2a). Many cholinergic fibers also entered the collagen gel that carried the cells. The increased density of cholinergic fibers was evident using both AChE histochemistry and AChE immunohistochemistry. Similar implants of collagen gels carrying the parent 3T3 cells into the striatum on the other side of the same host animals had no effects at all on the distribution or density of the surrounding striatal cholinergic innervation (Fig. 2b). These experiments thus demonstrate that adult cholinergic intrinsic neurons in striatum respond to added NGF and suggest that NGF sensitivity is a normal property of such neurons. In view of the low amounts of NGF mRNA present in the adult striatum (Whittemore et al. 1986), it appears that these neurons may not depend on continuous NGF support but may need it during development and perhaps after injury.

Using 3E cells to probe NGF sensitivity, we next wished to study the possible responses of cholinergic interneurons in cortex cerebri. The afferent cholinergic projections to frontoparietal cortex were therefore removed by stereotaxic injections of ibotenic acid into the nucleus basalis area to lesion the cellular origin of these projections. These lesions cause relatively complete cholinergic deafferentation as judged by the absence of cholinergic cells in anterior parts of nucleus basalis and a pronounced decrease of cholinergic nerve density in corresponding cortical areas. Small intracortical deposits of 3E or 3T3 cells were then made in the form of cell suspensions. Again, within a few weeks the density of cholinergic nerve terminals, as evaluated by both AChE histochemistry and immunohistochemistry, had increased significantly in a halo around the 3E cell implants, while no such effects were seen around the parent 3T3 cell implants. These experiments thus demonstrate that cortical cholinergic interneurons in adult animals are also responsive to NGF.

NGF-Secreting Genetically Engineered Cells Can Rescue Lesioned Cholinergic Neurons

One "classical" system in which to study cholinergic neurons and the effects of denervations, grafts, and trophic factors, as pioneered by Björklund, Gage and collaborators (Gage et al. 1984a, b, 1988; Gage and Björklund 1986), is the septo-hippocampal projection. We therefore also carried out fimbria-fornix lesions of this pathway in adult rats to cause cholinergic denervation of the hippocampal formation. Gels carrying 3E or 3T3 cells were then implanted into the lesioned area. Our preliminary observations in this system demonstrate a marked rescuing effect of the 3E cells on the septal cholinergic cell bodies as well as cholinergic reinnervation of hippocampal tissue, while the parent 3T3 cells were without effect (Strömberg et al., unpublished observations).

Similar results have recently been reported by Rosenberg et al. (1988), in which a retroviral vector was used to genetically modify fibroblasts to secrete NGF. Such cells were shown to prevent degeneration of cholinergic neurons after a fimbria-fornix lesion and to cause sprouting of cholinergic fibers towards the source of NGF. These results, and our own data, thus suggest that genetically engineered cells, secreting appropriate trophic factors, might be used to prevent degeneration and even to restore damaged fiber systems in disorders of the central nervous system.

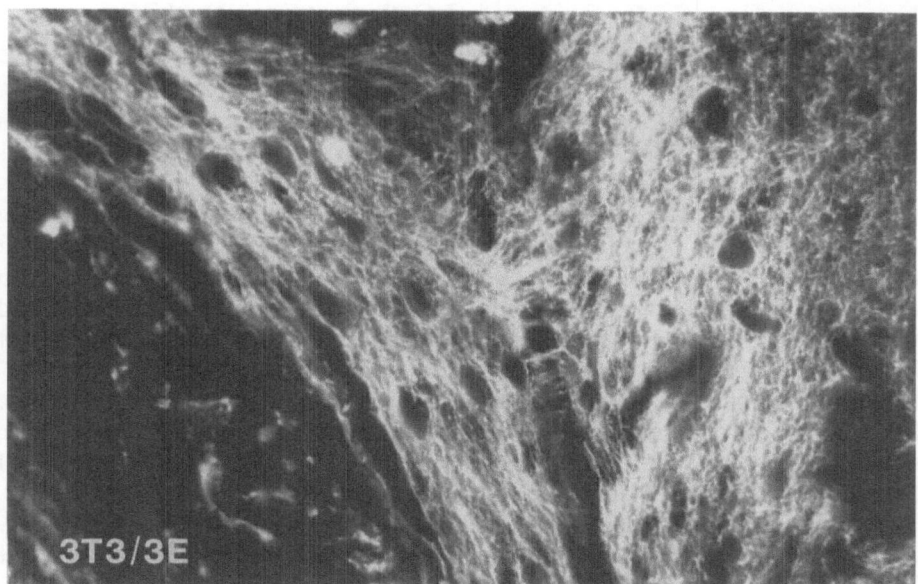

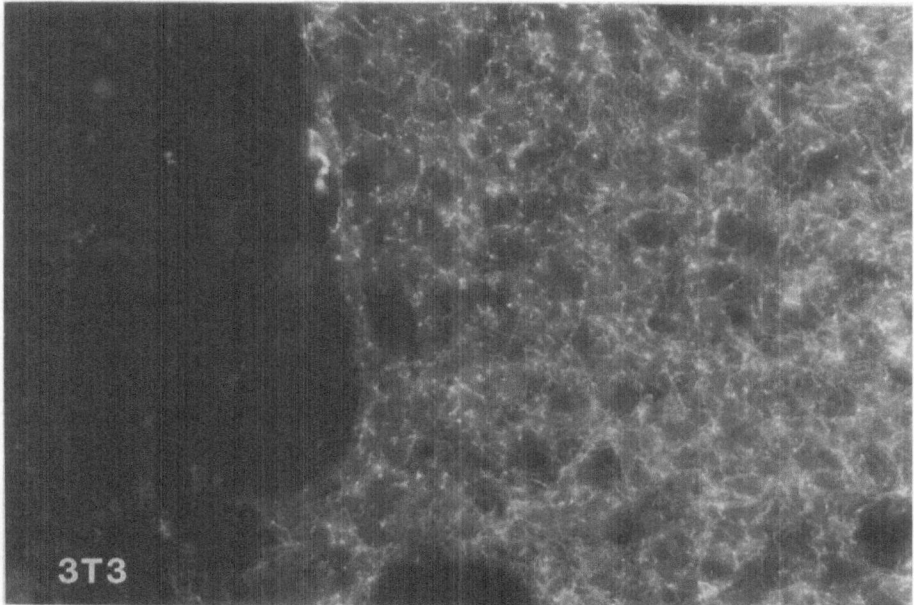

Fig. 2. Gels carrying 3E or 3T3 cells implanted into the intact striatum. Shown are the interfaces between gels *(left)* and host neuropil *(right)* stained with antibodies against AChE. Note the increased density of cholinergic terminals around the NGF-secreting 3E cells. Fluorescence microscopy, ×300. (From Ernfors et al., 1989a)

Cografting Genetically Engineered Cells that Secrete NGF Supports Grafts of Fetal Cholinergic Neurons

Another use for the NGF-secreting fibroblasts is to mix them with, or graft them adjacent to, other grafted cells that need NGF for optimal survival. One obvious example would be grafts of adrenal chromaffin cells which, as discussed above, survive better and undergo a much more complete neuronal transfer after grafting to striatum if NGF is delivered via a dialysis fiber (Strömberg et al. 1985). As a first test of the usefulness of 3E cells in cografting situations, we chose to mix them with cholinergic cell suspensions made from fetal basal forebrain. These mixtures, and similar mixtures using the 3T3 cells, were then injected into cortex cerebri of adult host rats in which the cholinergic projections to cortex had been lesioned by fimbria-fornix transection. The results were clearcut. In the presence of 3E cells a larger number of cholinergic neurons survived grafting and many more cholinergic nerve fibers were formed that radiated out into adjacent host cortical neuropil than in the presence of 3T3 cells (Ernfors et al. 1989a).

In a variation of the above experiment, fibroblasts and fetal cholinergic-rich cell suspensions were instead grafted separately but close to each other (about 1 mm) into the cholinergically denervated cortex cerebri. The findings were principally the same as with the mixtures, although now a directional (neurotropic, positive chemotactic) effect of NGF made by the 3E cells seemed evident. Thus, cholinergic fibers formed by the grafted cholinergic neurons seemed to grow preferentially towards the NGF source.

Taken together, the cografting experiments demonstrate the potential usefulness of cell lines secreting trophic factors as one way to support grafts of neurons and other cells to areas of the adult central nervous system where such factors might be absent, or present in too low concentrations for optimal taking of the grafted cells.

Concluding Remarks

In this chapter we have reviewed evidence for the manufacture of NGF in neurons in the adult central nervous system. We have also given several examples of which roles NGF might play in the central nervous system. Using genetically engineered cells that secrete considerable amounts of biologically active NGF as an in vivo probe for NGF sensitivity, we have shown that cholinergic interneurons in both cortex cerebri and striatum can respond to NGF and that injured cholinergic neurons projecting through of the septo-hippocampal pathway can be rescued by NGF. Cografting NGF-secreting and NGF-responsive cells has demonstrated improved survival and fiber formation of fetal cholinergic neurons. Our intraocular experiments have provided further evidence for the importance of NGF for cholinergic neurons in the basal forebrain and, in addition, have given strong support for an effect of NGF on noncholinergic neurons in the area. Moreover, a most interesting effect of NGF on development of cortex cerebri and hippocampus was suggested from intraocular grafting experiments.

The realization that NGF has a multitude of effects in the central nervous system, and the fact that effects are now being detected with many other growth factors and related compounds, together with the proven effects of cellular grafts in models of neurodegenerative diseases such as Parkinson's disease, provide hope for future treatment strategies. It is no longer inconceivable that effective treatments based on trophic factors, perhaps delivered by genetically engineered cells, grafts of fetal neurons and other cells, or combinations of trophic factors and grafts, may be developed for several severe neurological diseases such as, for example, Parkinson's disease, Alzheimer's disease, and spinal cord injury.

Acknowledgments. Supported by the Swedish Medical Research Council, the Swedish Natural Science Research Council, the Swedish Board for Technical Development, Magnus Bergvalls Stiftelse, Karolinska Institutets Fonder, Loo and Hans Ostermans Fond and US PHS grants NS09199 and AG04418

References

Alzheimer A (1907) Über eine eigenartige Erkrankung der Hirnrinde. Zentralbl Nervenheilkd Psychiatr 18:177−179

Arendt T, Bigl V, Arendt A, Tennstedt A (1985) Loss of neurons in the nucleus basalis of Meynert in Alzheimer's disease, paralysis agitans and Korsakoff's disease. Acta Neuropathol (Berl) 43:388−393

Ayer-LeLièvre C, Olson L, Ebendal T, Seiger Å, Persson H (1988) Expression of the beta-nerve growth factor gene in hippocampal neurons. Science 240:1339−1341

Ayer-LeLièvre C, Ebendal T, Olson L, Seiger Å, Persson H (1989) Detection of nerve growth factor and its mRNA by separate and combined immunohistochemistry and *in situ* hybridization in mouse salivary glands. Histochem J 21:1−7, 1989

Bartus RT, Dean RL, Beer B, Lippa AS (1982) The cholinergic hypothesis of geriatric memory dysfunction. Science 217:408−414

Bowen DM, Spillane JA, Curzon G, Meier-Ruge W, White P, Goodhardt MJ, Iwangoff P, Davison AN (1979) Accelerated ageing or selective neuronal loss as an important cause of dementia? Lancet 8106:11−14

Davies P, Maloney AJ (1976) Selective loss of central cholinergic neurons in Alzheimer's disease. Lancet 8000:1403

Ebendal T, Larhammar D, Persson H (1986) Structure and expression of the chicken beta nerve growth factor gene. EMBO J 5:1483−1487

Eriksdotter-Nilsson M, Skirboll S, Ebendal T, Hersh L, Grassi J, Massoulié J, Olson L (1989a) NGF treatment promotes development of basal forebrain tissue grafts in the anterior chamber of the eye. Exp Brain Res 74:89−98

Eriksdotter-Nilsson M, Skirboll S, Ebendal T, Olson L (1989b) Nerve growth factor can influence growth of cortex cerebri and hippocampus: evidence from intraocular grafts, Neuroscience (in press)

Ernfors P, Ebendal T, Olson L, Mouton P, Strömberg I, Persson H (1989a) A cell line producing recombinant NGF evokes growth responses in intrinsic and grafted central cholinergic neurons. Proc Natl Acad Sci USA (in press)

Ernfors P, Hallböök F, Ebendal T, Shooter EM, Persson H (1989b) Developmental and regional expression of beta-nerve growth factor receptor messenger RNA in the chick and rat. Neuron (in press)

Gage FH, Björklund A (1986) Cholinergic septal grafts into the hippocampal formation improve spatial learning and memory in aged rats by an atropine-sensitive mechanism. J Neurosci 6:2837−2847

Gage FH, Björklund A, Stenevi U (1984a) Denervation releases a neuronal survival factor in adult rat hippocampus. Nature 308:637−639

Gage FH, Björklund A, Stenevi U, Dunnett SB, Kelly PAT (1984b) Intrahippocampal septal grafts ameliorate learning impairments in aged rats. Science 225:533–536

Gage FH, Armstrong DM, Williams LR, Varon S (1988) Morphological response of axotomized septal neurons to nerve growth factor. J Comp Neurol 269:147–155

Korsching S, Auburger G, Heumann R, Scott J, Thoenen H (1985) Levels of nerve growth factor and its mRNA in the central nervous system of the rat correlate with cholinergic innervation. EMBO J 4:1389–1393

Lärkfors L, Ebendal T, Whittemore SR, Persson H, Hoffer B, Olson L (1987) Decreased level of nerve growth factor (NGF) and its messenger RNA in the aged rat brain. Mol Brain Res 3:55–60

Olson L, Seiger Å, Strömberg I (1983) Intraocular transplantation in rodents. A detailed account of the procedure and examples of its use in neurobiology with special reference to brain tissue grafting. In: Fedoroff S, Hertz L (eds) Advances in cellular neurobiology, vol 4. Academic, New York, pp 407–442

Olson L, Ernfors P, Ebendal T, Mouton P, Strömberg I, Persson H (1988) The establishment and use of stable cell lines that overexpress a transfected beta-nerve growth factor (NGF) gene: studies in vitro, in oculo and intracranially. Soc Neurosci Abstr 14 1:684

Pearson RCA, Sofroniew MW, Cuello AC, Powell TPS, Eckenstein F, Esiri MM, Wilcock GK (1983) Persistence of cholinergic neurons in the basal nucleus in a brain with senile dementia of the Alzheimer's type demonstrated by immunohistochemical staining for choline acetyltransferase. Brain Res 289:375–379

Perry EK, Perry RH, Blessed G, Tomlinson BE (1977) Necropsy evidence of central cholinergic deficits in senile dementia. Lancet 8004:189

Perry RH, Candy JM, Perry EK, Irving D, Blessed G, Fairbairn AF, Tomlinson BE (1982) Extensive loss of choline acetyltransferase activity is not reflected by neuronal loss in the nucleus of Meynert in Alzheimer's disease. Neurosci Lett 33:311–315

Rosenberg MB, Friedman T, Robertson RC, Tuszynski M, Wolff JA, Breakefield XO, Gage FH (1988) Grafting genetically modified cells to the damaged brain: restorative effects of NGF expression. Science 242:1575–1578

Shelton DL, Reichardt LF (1986) Studies on the expression of the beta-nerve growth factor (NGF) gene in the central nervous system: level and regional distribution of NGF mRNA suggest that NGF functions as a trophic factor for several distinct populations of neurons. Proc Natl Acad Sci USA 83:2714–2718

Strömberg I, Herrera-Marschitz M, Ungerstedt U, Ebendal T, Olson L (1985) Chronic implants of chromaffin tissue into the dopamine-denervated striatum. Effects of NGF on graft survival, fiber growth and rotational behavior. Exp Brain Res 60:335–349

Thoenen H, Bandtlow C, Heumann R (1987) The physiological function of nerve growth factor in the central nervous system: comparison with the periphery. Rev Physiol Biochem Pharmacol 109:145–178

White P, Hiley CR, Goodhardt MJ, Carrasco LH, Keet J, Williams EI, Bowen DM (1977) Neocortical cholinergic neurons in elderly people. Lancet 8013:668–671

Whitehouse P, Price DL, Struble RG, Clark AW, Coyle JT, de Long MR (1982) Alzheimer's disease and senile dementia: loss of neurons in the basal forebrain. Science 215:1237–1239

Whittemore SR, Ebendal T, Lärkfors L, Olson L, Seiger Å, Strömberg I, Persson H (1986) Developmental and regional expression of beta-nerve growth factor messenger RNA and protein in the rat central nervous system. Proc Natl Acad Sci USA 83:817–821

Short-Term Memory in Rodent Models of Ageing: Effects of Cortical Cholinergic Grafts

S. B. Dunnett

Summary

A paired-trial operant delayed response task has been developed to assess short-term memory function in rats. Aged rats manifest delay-dependent deficits in memory performance, whereas nucleus basalis lesions induce relatively less specific learning impairments. Both classes of deficit have been significantly ameliorated by cholinergic-rich transplants implanted in the neocortex.

Introduction

Disorders of memory and new learning are the cardinal symptoms of dementia and often provide the earliest signs of the insidious onset of Alzheimer's disease (Sjögren 1952). All aspects of memory performance are disrupted in Alzheimer's disease, such that impairments are seen in both short-term and long-term, episodic and semantic memory (Morris and Kopelman 1986).

It has been suggested that the memory impairments observed in both normal and pathological ageing may be attributable to a decline in the function of the cholinergic innervation of the neocortex that originates in the subcortical basal nucleus of Meynert (Bartus et al. 1982; Coyle et al. 1983; Drachman and Sahakian 1980; Höhmann et al. 1988) Not only do patients dying with Alzheimer's disease manifest substantial reductions in choline acetyltransferase (ChAT) activity in the neocortex (Bowen et al. 1976; Davies and Maloney 1976; Perry et al. 1977) and loss or atrophy of the subcortical cholinergic neurons of the basal forebrain (Whitehouse et al. 1982; Pearson et al. 1983; Etienne et al. 1986), but these post-mortem indices are seen to correlate well with mental test scores collected from the patients while alive (Perry et al. 1978; Wilcock et al. 1982). Although it is likely that the cholinergic decline is not the primary neuropathological change in Alzheimer's disease but a secondary retrograde consequence of cortical cell loss (Pearson et al. 1985; Pearson and Powell 1987; Hardy et al. 1986; Mann 1988), the "cholinergic hypothesis" has stimulated considerable research on the functional contribution of central cholinergic systems to cognitive and mnemonic capacities in man and experimental animals. An underlying theme in this research is the prospect of designing rational replacement therapies to ameliorate the symptoms of dementia, to the extent that cholinergic decline contributes to those symptoms.

Cholinergic decline and replacement therapies have been modelled in experimental animals using anticholinergic drugs, lesions of the nucleus basalis magnocellularis (NBM; the rodent equivalent of the basal nucleus of Meynert), and the ageing process itself (for reviews see Bartus et al. 1982, 1983; Coyle et al. 1983; Fisher and Hanin 1986; Price 1986; Wenk and Olton 1987; Dunnett and Barth 1989; Smith 1988; Collerton 1986; Hagan and Morris 1988). All these treatments have been found to disrupt performance of rats on tasks dependent upon memory, such as passive avoidance retention of an aversive experience, active avoidance, spatial maze learning, and learning of complex discriminations. However, deficits in the majority of tasks employed in these studies may not be attributable specifically to a disruption of memory capacity, and an equally cogent case can be made for changes in the animals' sensitivity or attention to relevant environmental stimuli, or other activational or motivational changes (Collerton 1986; Warburton 1974; Heise 1975; Everitt et al. 1987; Hagan and Morris 1988). Our recent research interest has, therefore, focussed on the development of specific tests of memory function in rats that are not confounded by other explanations of the behavioural impairment. The present review concentrates on recent studies with an operant delayed response task to measure short-term memory capacity of rats, which has been used to assess deficits in several rodent models of ageing, and the capacity of cortical cholinergic transplants to ameliorate those deficits.

An Operant Delayed Response Task for Rats

Discrete-trial delayed response tasks provide a powerful means to assess the cognitive capacity of animals (including humans). Delayed responding typically involves the provision of information to an animal (e.g. the location of hidden food), which it can use subsequently in order to achieve reward (e.g. the selection of the appropriate location or stimuli associated with the food). As reviewed by Heise (1975), the use of discrete trials enables the controlled presentation of stimuli to animals which they can use to determine their subsequent response choices.

The most widely used delayed response tasks in primate studies are delayed matching and delayed non-matching to sample. The design and rationale of such tasks are illustrated in Figure 1a. The monkey is initially shown an object (or other sample stimulus) on the sample trial, and then after a variable delay interval is given a choice trial on which it has to choose between the sample and a novel object. Food reward is then hidden under the sample object in matching tasks or under the novel object in the non-matching version. Such tasks require that the animals employ some mnemonic strategy to hold the initial information until they have the opportunity for responding, and the slope of the delay−performance function provides an index of the rate of forgetting from short-term memory (Fig. 1b).

The power of delayed matching and non-matching tasks lies in the identification of specific changes in the rate of forgetting of information from short-term memory by a change of slope of the delay−performance function (Fig. 1c). By contrast, a non-specific impairment (e.g. one involving sensory, motor, or motiva-

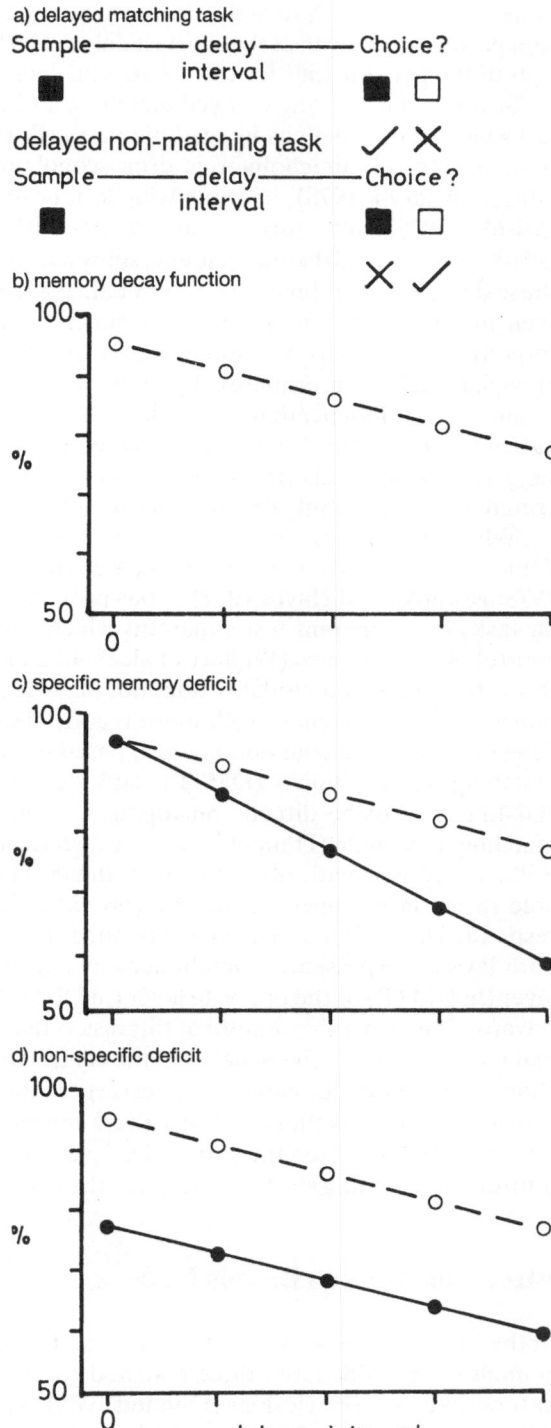

Fig. 1a–d. Schematic illustration of the delayed matching and non-matching to sample paradigm. **a** On each trial a sample stimulus *(filled square)* is presented, and then after a variable interval delay the animal makes a choice response between the sample and a different (open square) stimulus. Choice of the sample is rewarded in delayed matching paradigms; choice of the other stimulus is rewarded in non-matching paradigms. **b** The rate of forgetting from short-term memory is represented by the slope of the delay-performance function. **c** A specific short-term memory impairment leaves performance normal at the shortest delays but involves a progressively greater deterioration in performance at longer delays, reflected by an increased slope of the forgetting function *(solid line)*. **d** A non-mnemonic impairment in task performance is associated with comparable disruption of performance at all delays *(solid line)*. The plots in **b–d** are idealized and do not represent actual data

tional changes, or failure of the animal to attend to the relevant stimuli) would disrupt performance even at the shortest delays and hence is reflected by a downward shift of the performance function across all delays (Fig. 1d).

Several studies using delayed matching and delayed non-matching to sample tasks have shown deficits in aged monkeys (Bartus et al. 1978), young monkeys treated with the anticholinergic drug scopolamine (Bartus and Johnson 1976; Glick and Jarvik 1978), and following lesions to the hippocampus and amygdala (Mishkin 1978; Zola-Morgan et al. 1982) or NBM (Aigner et al. 1987; Ridley et al. 1986). Similarly, Alzheimer patients show marked impairments on tasks based on those designed for primates (Flicker et al. 1984; Sahakian et al. 1988). The deficit seen in Alzheimer patients is delay-dependent, reflecting an increased rate of forgetting, and this is well modelled by the studies of ageing and hippocampal/ amygdala lesions in primates. By contrast, the pattern of impairment has been found to be independent of delays in models involving NBM lesions or anticholinergic drug treatments (Aigner et al. 1987; Glick and Jarvik 1978), suggesting in these cases that the treatments affected the animals attention or discrimination of relevant stimuli rather than short-term memory capacity per se.

Whereas delayed matching and non-matching performance has been demonstrated in rats in open mazes (Aggleton 1985; Alexinsky and Chapouthier 1978; Rothblat and Hayes 1987), it has proved more difficult to train rats on similar tasks in an operant test apparatus which would enable greater experimental control of performance (Wallace et al. 1980; Dunnett 1985). A possible reason for this is that the visual modality does not provide salient stimuli for rats: rats learn olfactory or spatial cues much more readily (Hodos 1970). We have, therefore, developed a paired-trial operant task based on spatial sample stimuli − "delayed matching to position" (DMTP) and "delayed non-matching to position" (DNMTP) − to be directly analogous to primate delayed matching and non-matching to sample (Dunnett 1985; see also Kesner et al. 1981). The DMTP task is illustrated schematically in Figure 2. In DMTP and DNMTP one of two retractable levers in the operant chamber provides the sample to which the rat must respond. Then, after a variable delay, the rat is confronted with the choice phase: both levers are presented simultaneously and the rat must respond to the same lever (in DMTP) or the opposite lever (in DNMTP) as the previous sample to gain reward. The distinctive feature of this task is that during the delay interval the animal must respond to the panel over the central food well to obtain insertion of the choice levers into the chamber. Panel responding (on a variable interval schedule) provides a supplementary distractor task during the delay interval and keeps the rat centralized between the two choice levers, thereby preventing the adoption of position or orienting strategies to solve the task without recourse to memory.

Effects of Nucleus Basalis Lesions

In the first study of operant DMTP in rats, following pretraining on the task, the animals were divided into three matched groups to receive either control, NBM, or fimbria-fornix (FF) lesions (Dunnett 1985). The NBM lesions were made by the stereotaxic injection of 0.4 µl 0.6 M ibotenic acid into the vicinity of the magnocel-

Fig. 2. Schematic illustration of the delayed matching to position (DMTP) task. *Above* is shown the side wall of the operant chamber, with a hinged square perspex panel covering the central food well and a retractable response lever on either side. Illumination is provided by a house light in the ceiling and a panel light behind the perspex panel. Additional stimulus lights above the panel and each response lever are not used at any stage of the task. *Below* is shown each stage of a trial viewed from above, with each response made by the rat indicated by an *arrow*. On each trial the side of the sample lever and the variable interval delay (*VI*, range 0–16 s, 0–24 s, or 0–32 s in the different experiments) are selected at random. The delayed non-matching to position (DNMTP) task is identical apart from the one difference that the "error" contingencies follow a matching response and the "correct" contingencies follow a non-matching response. (Redrawn from Dunnett 1987)

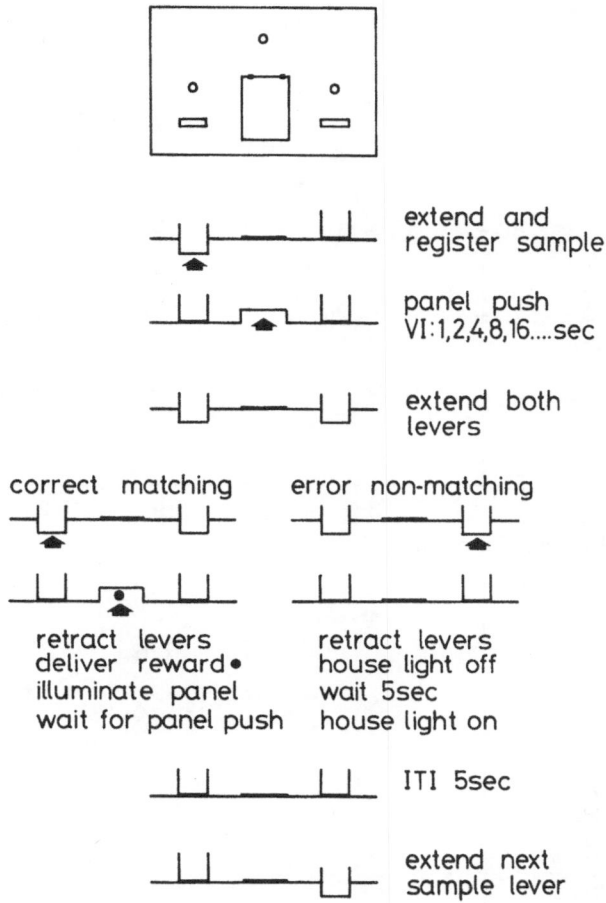

extend and register sample

panel push VI:1,2,4,8,16....sec

extend both levers

correct matching   error non-matching

retract levers   retract levers
deliver reward •   house light off
illuminate panel   wait 5sec
wait for panel push   house light on

ITI 5sec

extend next sample lever

lular cholinergic cells in the ventral pallidum/substantia innominata. The FF lesions were made by aspirative transection of the fibre bundle under visual guidance. These lesions induced extensive cholinergic deafferentation of the dorsolateral neocortex and of the hippocampus, respectively, as assessed by acetylcholinesterase (AChE) histochemistry. Training on the delayed matching task recommenced 5 weeks later.

As shown in Figure 3a, both NBM and FF lesions impaired post-operative retention of task performance. Although the deficit was greater in the NBM group in the first few days of post-operative testing, these rats rapidly relearned the task up to control levels within approximately 15 days. By contrast, although the FF lesions initially induced a somewhat less severe impairment, that impairment did not decline with further training. Inspection of performance at different delay intervals indicated that the deficit in the rats with FF lesions was delay-dependent, whereas the deficit induced by NBM lesions was constant across all delays includ-

a) Delayed matching

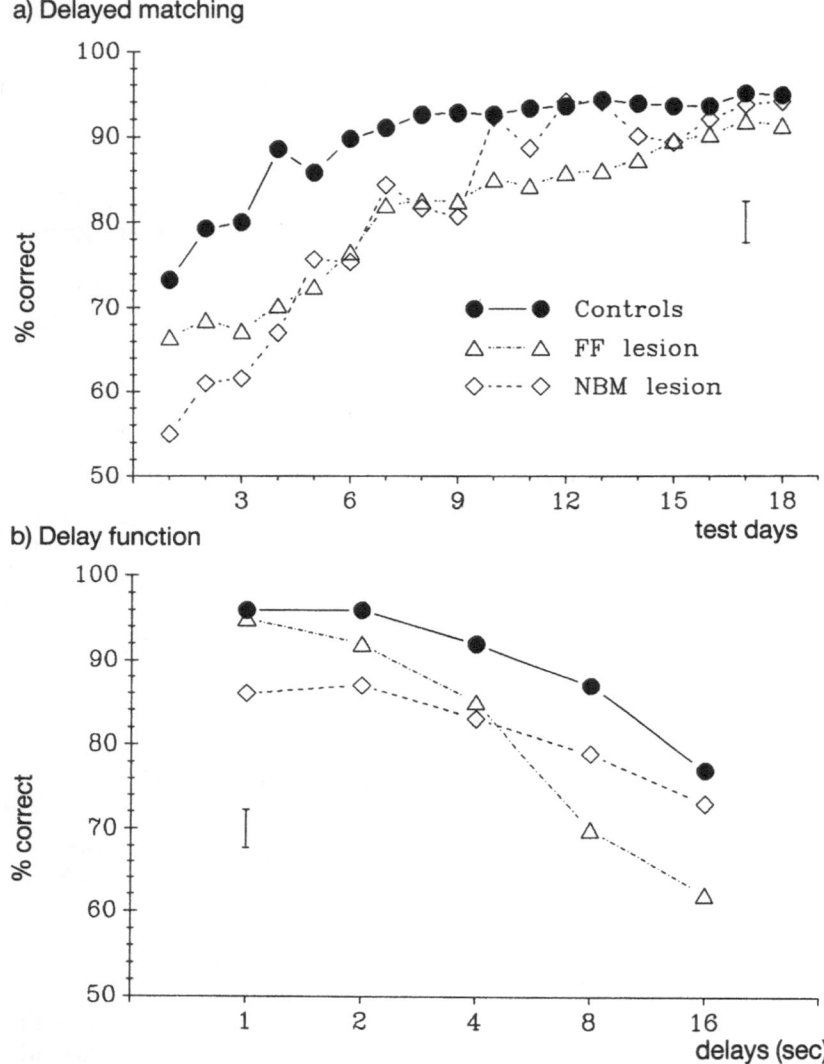

Fig. 3a—b. Effects of ibotenic acid nucleus basalis *(NBM)* and aspirative fimbria-fornix *(FF)* lesions on delayed matching (DMTP) performance in young rats. **a** Performance on the first 18 days of testing following lesion surgery collapsed accross different delay intervals. **b** Performance at each delay (1−16 s) collapsed across the 18 days of testing. Vertical bars indicate 2 SEM (Data from Dunnett 1985)

ing the shortest (see Fig. 3b). Thus, the FF lesions appeared to produce a specific deficit in the rat's ability to retain information in short-term ("working") memory, akin to the effects in monkeys with hippocampal lesions (Mishkin 1978; Zola-Morgan et al. 1982) and Alzheimer patients (Sahakian et al. 1988) trained on similar tasks, and the effects in rats with similar lesions on radial maze tasks (Olton et al. 1978). By contrast, the NBM lesions appeared to induce a generalised disruption of pre-operatively trained performance from which the animals recovered with further training. So although the NBM lesioned rats were impaired on the short-term memory task, the delay-independent nature of their impairments does not permit the conclusion that the deficit was explicitly mnemonic.

Cholinergic Transplants in Lesioned Rats

To what extent can neural transplants provide a cortical cholinergic replacement therapy for various deficits associated with NBM lesions? Several studies have previously demonstrated the capacity of cholinergic grafts in the hippocampus to ameliorate deficits associated with lesions in the hippocampal circuitry (Dunnett et al. 1982; Nilsson et al. 1987). We have therefore tried similar grafts of embryonic cholinergic neurons to the neocortex of rats with NBM lesions. The demonstration of functional recovery associated with a graft-derived cholinergic reinnervation of the neocortex would strengthen the hypothesis that those particular deficits were indeed attributable to cortical deafferentation (Fine et al. 1985a; Dunnett et al. 1985).

    Cortical cholinergic transplants have been made by stereotaxic injection of dissociated cell suspensions prepared by standard procedures which have been described in detail elsewhere (Björklund et al. 1983; Dunnett et al. 1986) and illustrated schematically in Figure 4. Such grafts are rich in cholinergic neurons which can be visualized by both AChE histochemistry and ChAT immunohistochemistry (Fine et al. 1985b; Dunnett et al. 1986). The grafts give rise to a dense outgrowth of AChE positive fibres over 1–2 mm radius in the host neocortex (see Fig. 5). Biochemical assay of host neocortex in the area of the outgrowth proximal to the

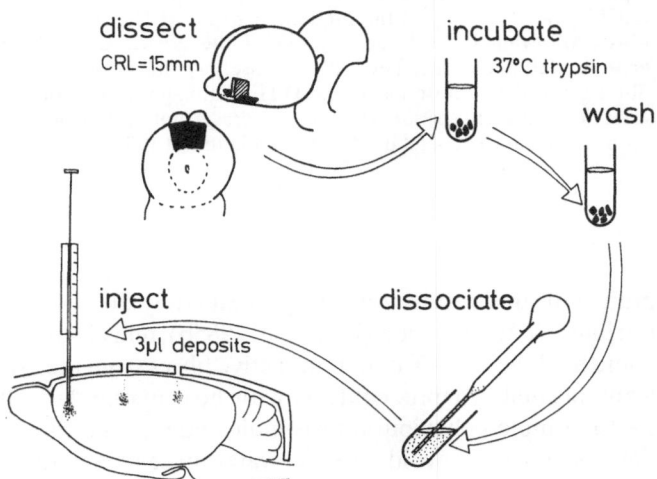

Fig. 4: Schematic diagram of the transplantation procedure. Ventral forebrain tissue, rich in the developing cholinergic cells of the septum and diagonal band, is dissected from E15–16 embryos, incubated in trypsin, washed, and mechanically dissociated to form a dense cell suspension. One or several 3 µl aliquots of the suspension are stereotaxically injected via a 10 µl microsyringe into the neocortex of the host animal. (Redrawn from Dunnett 1987)

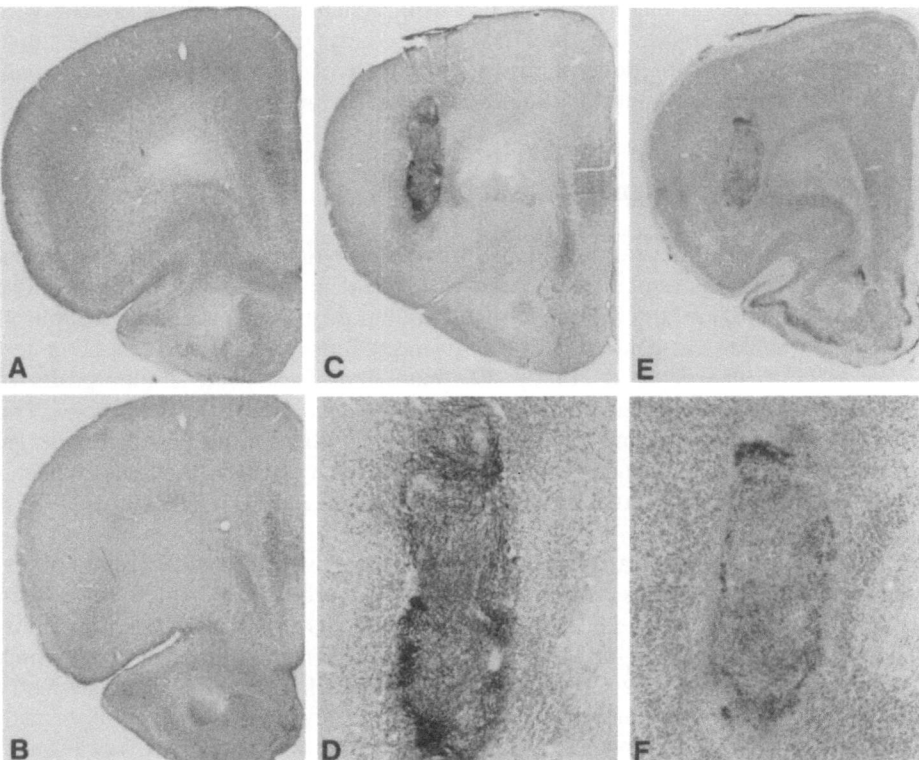

Fig. 5A–F. Histology of cholinergic grafts in the frontal neocortex. **A** Acetylcholinesterase (AChE) fibre-staining in the intact neocortex. **B** Deafferentation of cortical AChE-positive fibres following ibotenic acid lesion of the NBM. **C** Cholinergic graft implanted in the deafferented neocortex. The graft is densely AChE positive and gives rise to a halo of AChE fibres growing into host neocortex. **D** Higher magnification of graft shown in c. **E, F** The same graft in an adjacent section, stained with cresyl violet, showing healthy cells in the graft and the absence of a marked glial border. (From Dunnett 1987)

graft indicates restoration of approximately 50% of the loss of ChAT activity attributable to the lesion (Fine et al. 1985b), and electron microscopic studies have confirmed that ChAT immunoreactive fibres from the grafts establish morpholog-ically normal synaptic contacts with host pyramidal cells (Clarke and Dunnett 1986). Simple behavioural tests indicated that the cholinergic grafts ameliorated deficits in passive avoidance and water maze navigation in rats with NBM lesions (Dunnett et al. 1985; Fine et al. 1985a).

What effects do the grafts have on DMTP performance of NBM lesioned rats? In one study 34 rats were pretrained on the DMTP task, following which 11 received NBM lesions alone and a further 8 received NBM lesions plus cholinergic grafts into six sites in the deafferented neocortex (Dunnett 1987). Three months later the rats were all retested on the task. As shown in Figure 6, the performance of the rats with NBM lesions in the retention tests was below that of the control group, although in this experiment the lesion-induced deficit was less severe than in the previous experiment. This is perhaps due to the fact that the rats were tested after a longer post-operative interval. Nevertheless, the lesioned rats with additional cholinergic grafts showed a substantial improvement in performance on the retention tests at least up to control levels. This supports the hypothesis that the

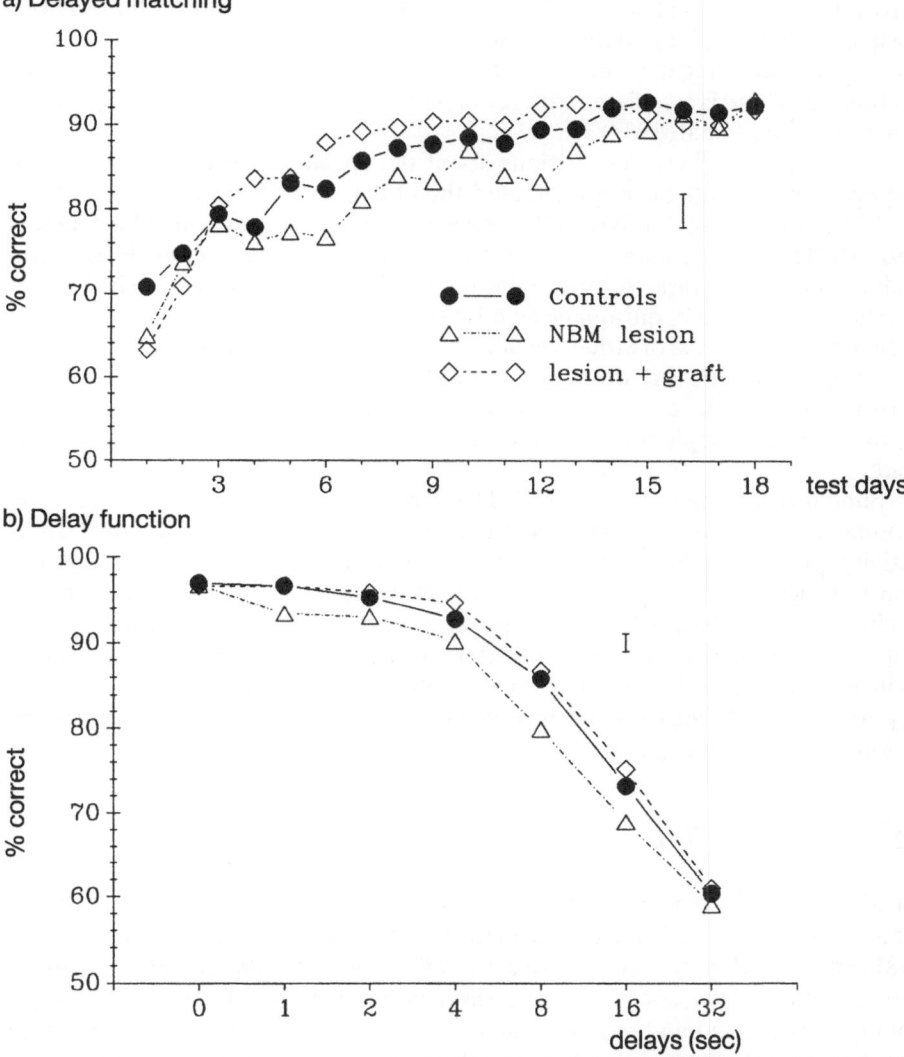

a) Delayed matching

b) Delay function

Fig. 6a, b. Effects of cholinergic-rich grafts implanted bilaterally into the neocortex of rats with ibotenic acid NBM lesions on delayed matching (DMTP) performance in young rats. **a** Performance on the first 18 days of testing following lesion surgery, collapsed across different delay intervals. **b** Performance at each delay (0–32 s), collapsed across the 18 days of testing. Vertical bars indicate 2 SEM (Data from Dunnett 1987)

post-operative deficits induced by the NBM lesions are at least in part attributable to cholinergic deafferentation of the neocortex.

Specificity of Ibotenic Acid Lesions

The major interpretative difficulty with these studies is that ibotenic acid lesions of the NBM induce widespread non-cholinergic neuropathological damage and associated neurological deficits (Dunnett et al. 1987; Everitt et al. 1987). By contrast, quisqualic acid, although not a specific cholinergic neurotoxin, is more selective against the magnocellular cholinergic neurons in the basal forebrain than ibotenic acid, and is associated with fewer neurological and neuropathological side effects (Dunnett et al. 1987).

In the light of these reservations about the specificity of ibotenic acid, subsequent experiments have investigated the effects of bilateral injections of 1 µl 0.12 M quisqualic acid into the NBM on acquisition and retention of both DMTP and DNMTP tasks (Dunnett et al. 1989). In one experiment, animals were pretrained on the two tasks prior to allocation to balanced groups for NBM lesion. As shown in Figure 7, quisqualic acid lesions of the NBM induced no detectable deficit in performance of either version at any delay, even during the first week of retention testing. Post-mortem biochemical analysis has confirmed a 55% loss of ChAT activity in the cortex of these NBM lesioned rats, which is at least as great as the extent of depletion observed in behavioural studies employing ibotenic acid.

Similar results have been observed by Etherington et al. (1987). In that study, ibotenic acid lesions of the NBM (associated with 17% decline in cortical ChAT activity) produced marked impairments in an operant paired-trial delayed alternation task, whereas quisqualic acid lesions of the same site (associated with 55% decline in cortical ChAT activity) produced no significant impairments. Consequently, the absence of any deficit after quisqualate lesions cannot be attributed to inadequacy of the lesions. These observations place a severe constraint on any hypothesis postulating a specific involvement of NBM systems in either short-term or longer term memory capacities.

Effects in Aged Rats

Aged rats show deficits in a wide variety of tasks dependent on memory performance, including passive avoidance (Bartus et al. 1983; Kubanis and Zornetzer 1981) and spatial maze learning (Barnes 1979; Gage et al. 1984b). We have therefore investigated the specific short-term memory capacity of separate groups of young, middle-aged and old rats (aged 6, 15 and 24 months, respectively, at the commencement of testing) on both matching and alternation versions of the operant delayed response task (Dunnett et al. 1988). Performance during acquisition of the delayed non-matching task is shown in Figure 8a. The old group of rats showed substantial and significant deficits on both versions and never reached the same level of asymptotic performance as young animals.

The reason for this age-related deficit in delayed responding is revealed when the asymptotic performance of each group is analysed in terms of performance at each delay interval (Fig. 8b). All groups showed near perfect performance at the shortest delays of 0, 1 or 2 seconds and a progressive decline in performance at longer delays. However, the rate of decline is greater in the 24-month group than

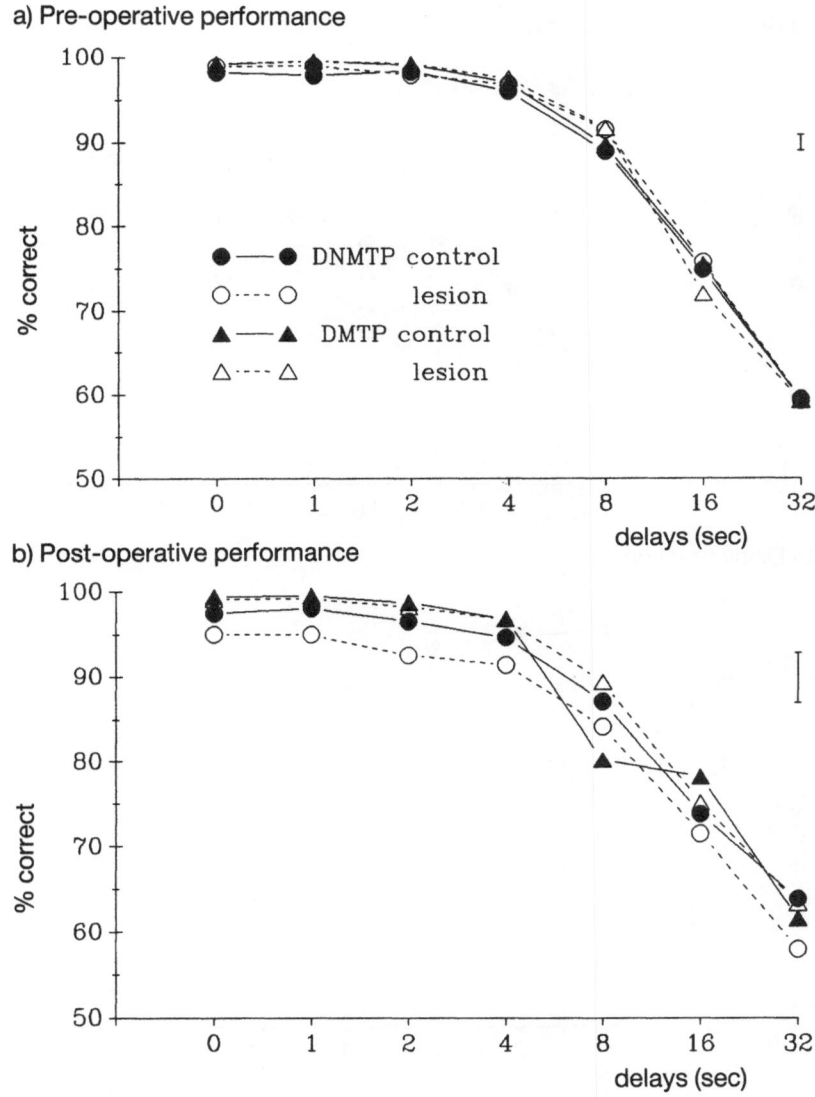

Fig. 7a, b. Effects of quisqualic acid NBM lesions on delayed matching (DMTP) and non-matching (DNMTP) performance in young rats. **a** Pre-operative performance at each delay (0–32 s) on the last seven days training prior to lesion surgery. **b** Post-operative performance at each delay on the first seven days training after lesion surgery. Vertical bars indicate 2 SEM (Data from Dunnett et al. 1989)

in the 6-month group on both versions of the task, suggesting a higher rate of forgetting of the sample information from short-term memory. The middle-aged, 15-month, group was unimpaired on the delayed alternation version (Fig. 8b), although significant deficits were apparent by this age on the delayed matching version (Dunnet et al. 1988). These data indicate that old rats do have specific delay-dependent impairments in their short-term memory capacity.

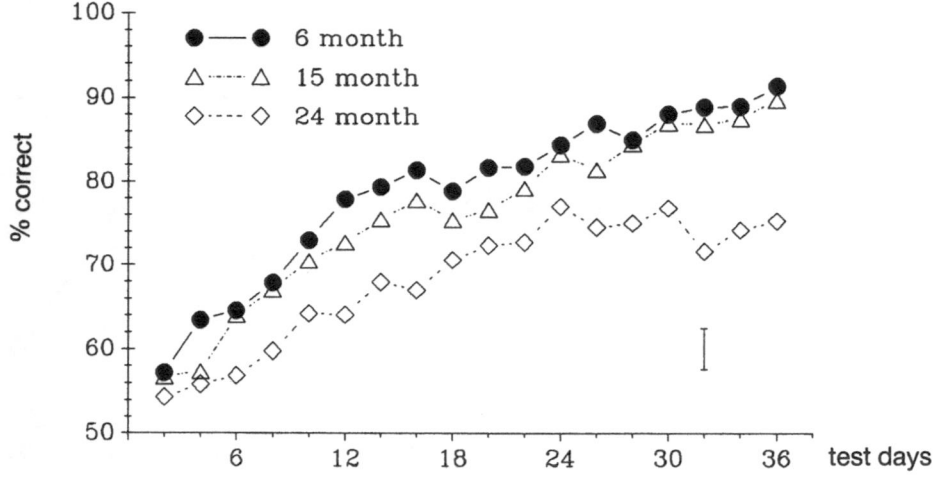

a) Delayed non-matching

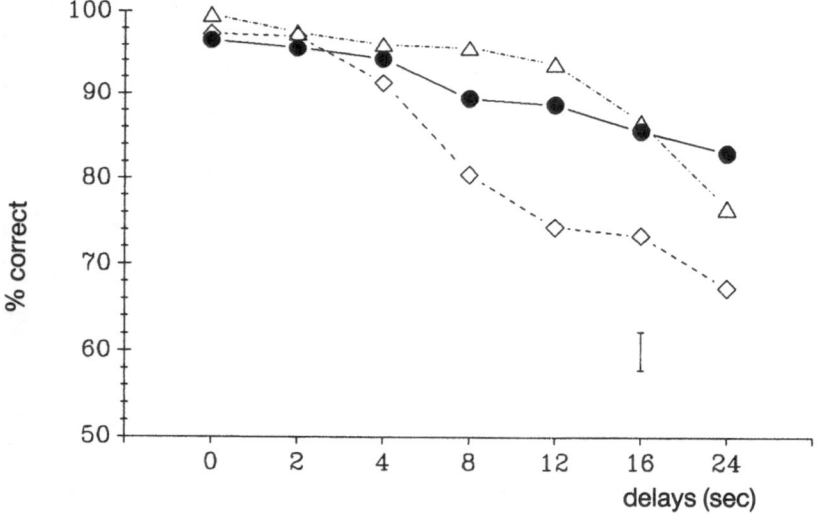

b) Delay function

Fig. 8a, b. Acquisition of the delayed non-matching (DNMTP) task by rats of 6, 15 and 24 months of age. **a** Performance on the first 36 days of training, collapsed across different delay intervals. **b** Performance at each delay (0–24 s) on the last seven days of training, at which time performance had attained asymptotic levels. Vertical bars indicate 2 SEM (Data from Dunnett et al. 1988)

It has been suggested that if declining memory capacity in ageing is attributable to a decline in cholinergic activation in the cortex then cholinergic replacement therapies might be effective (Bartus et al. 1982, 1983). We therefore administered a range of doses of the cholinergic agonist arecoline and the cholinesterase inhibitor physostigmine to the rats. Bartus et al. (1983) have suggested that both of these drugs can improve aged rats' performance on passive avoidance tests. However, we were unable to find any dose of either drug that produced a general improvement of performance in all groups or indeed any selective benefit to the aged rats. These pharmacological treatments do not, therefore, support the hypothesis that generalized pharmacological activation of cholinergic systems can ameliorate short-term memory deficits in aged rats.

Cholinergic Grafts in Aged Rats

Although cholinergic replacement by pharmacological therapy was ineffective in the old rats, the synaptic reinnervation and greater integration provided by cholinergic grafts in young animals warranted investigation of similar grafts in the aged brain. Cholinergic grafts implanted in the hippocampus of aged rats can ameliorate deficits in spatial maze tasks (Gage et al. 1984a; Gage and Björklund 1986), although the effects of cortical placements have not previously been investigated. We have therefore compared the effects of cholinergic grafts in the cortex and in the hippocampus on the ability of old rats to learn DNMTP tasks (Dunnett et al. 1989). As shown in Figure 9a, the deficits in learning the task by old rats were replicated with both graft sites. Moreover, cholinergic grafts in either site provided a modest but significant improvement in asymptotic performance. Analysis of delay-dependent performance over the last 7 days of training, when the rats were responding at asymptotic levels, showed again that the old rats had learned the task and were unimpaired at the short delays when memory load was minimal. By contrast, the increased rate of forgetting by the old rats at intermediate delays when the memory load was moderate was significantly improved by the grafts (Fig. 9b). These observations suggest that the cholinergic innervation of both cortical and hippocampal systems may be implicated in the age-related decline in short-term memory.

Reconciliation of Ageing and NBM Lesion Effects

The different experiments involving pharmacological treatments and cholinergic grafts in aged rats and rats with NBM lesions appear to yield contradictory conclusions. On the one hand, the generalized disruption of performance induced by scopolamine (Dunnett 1985; Dunnett et al. 1989) and ibotenic acid lesions of the NBM appear to suggest that any impairment in task performance is not specifically related to short-term memory capacity. Moreover, the effects of quisqualic acid lesions suggest that the disruption of performance induced by ibotenic acid lesions of the NBM is related to non-specific basal forebrain damage, independent of influences on the magnocellular population of cholinergic neurons. On the other

a) Delayed non-matching

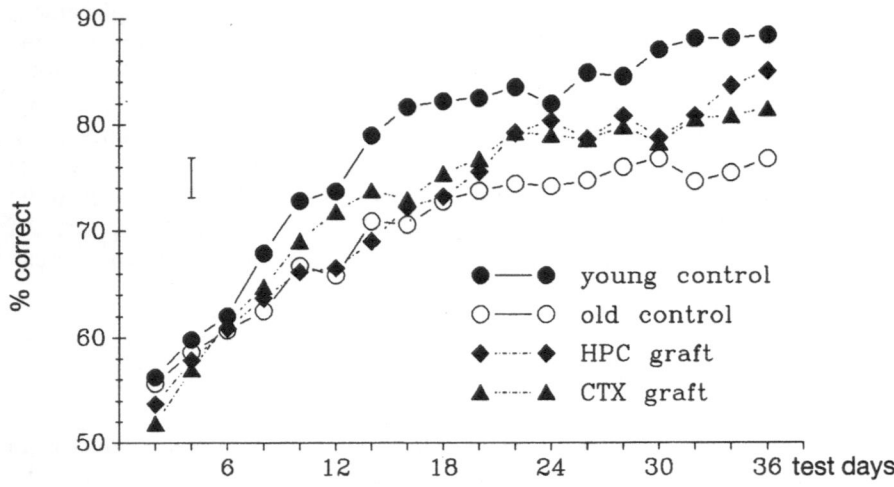

b) Delay function

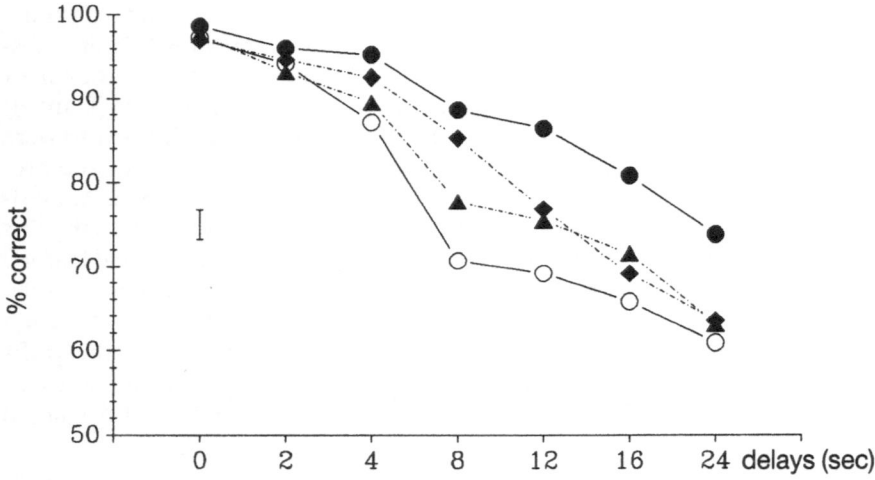

Fig. 9a, b. Effects of cholinergic grafts implanted in the neocortex *(CTX)* or hippocampus *(HPC)* on acquisition of the delayed non-matching (DNMTP) task by old rats. **a** Performance on the first 36 days of training, collapsed across different delay intervals. **b** Performance at each delay (0–24 s) on the last seven days of training, at which time performance had attained asymptotic levels. Vertical bars indicate 2 SEM (Redrawn from Dunnett et al. 1989)

hand, aged rats do show specific impairments in short-term memory capacity, and the different deficits of aged rats and ibotenic acid lesioned rats are both amenable to amelioration by cortical cholinergic grafts.

These discrepancies can be resolved by consideration of the topographic organization of basal forebrain cholinergic systems. The deficits of aged rats on spatial navigation tasks are associated with changes in metabolic activity in the hippocampus and frontal cortex (Gage et al. 1984c), and grafts in either of these two sites produced partial improvement in the aged rat deficits in the delayed response task. NBM lesions induce learning and long-term retention deficits to a varying degree of specificity depending on the choice of neurotoxin, rather than short-term memory disturbance. However, this lesion deafferents the dorsolateral neocortex of rats, including the parietal association areas, but spares the innervation to both prefrontal cortical and hippocampal sites. By contrast, the effective grafts in the ageing study included placements in both medial and orbital prefrontal sites, the cholinergic innervation of which arises from neurons predominantly located in the diagonal band of Broca rather than in the NBM proper (Luiten et al. 1987).

Thus, the search for a unitary or unique function associated with basal forebrain-cortical cholinergic systems is naive. The profile of deficits induced by basal forebrain lesions will depend on both the precise pattern of cortical deafferentation and the degree of cholinergic specificity of the neurotoxin. Similarly, the profile and degree of recovery induced by grafts will depend on the specificity of graft tissues, the neurochemical as well as topographical pattern of reinnervation, and the selection of behavioural functions that are relevant to the locus of deafferentation and graft placement.

Acknowledgements. These studies were supported by grants from the Mental Health Foundation and the Medical Research Council. I am indebted to my collaborators and colleagues: A Björklund, FH Gage, IQ Whishaw, SD Iversen, JL Evenden, TW Robbins, A Fine, DC Rogers, F Badman, GH Jones and ST Bunch.

References

Aggleton JP (1985) One-trial object recognition by rats. Q J Exp Psychol [B] 37:279–294

Aigner TG, Mitchell SJ, Aggleton JP, DeLong MR, Struble RG, Price DL, Wenk GL, Mishkin M (1987) Effects of scopolamine and physostigmine on recognition memory in monkeys with ibotenic-acid lesions of the nucleus basalis of Meynert. Psychopharmacology (Berlin) 92:292–300

Alexinsky T, Chapouthier G (1978) A new behavioral model for studying delayed response in rats. Behav Biol 24:442–456

Barnes CA (1979) Memory deficits associated with senescence: a neurophysiological and behavioral study in the rat. J Comp Physiol Psychol 93:74–104

Bartus RT, Johnson HR (1976) Short-term memory in the rhesus monkey: disruption from the anti-cholinergic scopolamine. Physiol Behav 11:571–575

Bartus RT, Fleming D, Johnson HR (1978) Aging in the rhesus monkey: debilitating effects on short-term memory. J Gerontol 33:858–871

Bartus RT, Dean RL, Beer B, Lippa AS (1982) The cholinergic hypothesis of geriatric memory dysfunction. Science 217:408–417

Bartus RT, Flicker C, Dean RL (1983) Logical principles for the development of animal models of age-related memory impairments. In: Crook T, Ferris S, Bartus R (eds) Assessment in geriatric psychopharmacology. Powley, New Canaan, pp 263–299

Björklund A, Stenevi U, Schmidt RH, Dunnett SB, Gage FH (1983) Intracerebral grafting of neuronal cell suspensions. I. Introduction and general methods of preparation. Acta Physiol Scand [Suppl] 522:1–7

Bowen DM, Smith CB, White P, Davison AN (1976) Neurotransmitter-related enzymes and indices of hypoxia in senile dementia and other abiotrophies. Brain 99:459–496

Clarke DJ, Dunnett SB (1986) Ultrastructural organization of choline acetyltransferase-immunoreactive fibres innervating the neocortex from embryonic ventral forebrain grafts. J Comp Neurol 250:192–205

Collerton D (1986) Cholinergic function and intellectual decline in Alzheimer's disease. Neuroscience 19:1–28

Coyle JT, Price DL, DeLong MR (1983) Alzheimer's disease: a disorder of cortical cholinergic innervation. Science 219:1184–1190

Davies P, Maloney AJF (1976) Selective loss of central cholinergic neurons in Alzheimer's disease. Lancet II:1403

Drachman DA, Sahakian BJ (1980) Memory, aging and pharmacosystems. In: Stein D (ed) The psychobiology of aging: Problems and perspectives. Elsevier, Amsterdam, pp 347–368

Dunnett SB (1985) Comparative effects of cholinergic drugs and lesions of nucleus basalis or fimbria-fornix on delayed matching in rats. Psychopharmacology (Berlin) 87:357–363

Dunnett SB (1987) Anatomical and behavioral consequences of cholinergic-rich grafts to the neocortex of rats with lesions of the nucleus basalis magnocellularis. Ann NY Acad Sci 495:415–429

Dunnett SB, Barth TM (1989) Animal models of Alzheimer's disease and dementia (with an emphasis on cortical cholinergic systems). In: Willner P (ed) Behavioural models in psychopharmacology. Cambridge University Press, Cambridge (in press)

Dunnett SB, Low WC, Iversen SD, Stenevi U, Björklund A (1982) Septal trasplants restore maze learning in rats with fornix-fimbria lesions. Brain Res 251:335–348

Dunnett SB, Toniolo G, Fine A, Ryan CN, Björklund A, Iversen SD (1985) Transplantation of embryonic ventral forebrain neurons to the neocortex of rats with lesions of nucleus basalis magnocellularis. II. Sensorimotor and learning impairments. Neuroscience 16:787–797

Dunnett SB, Wishaw IQ, Bunch ST, Fine A (1986) Acetylcholine-rich neuronal grafts in the forebrain of rats: effects of environmental enrichment, neonatal noradrenaline depletion, host transplantation site and regional source of embryonic donor cells on graft size and acetyl-cholinesterase-positive fibre outgrowth. Brain Res 378:357–373

Dunnett SB, Wishaw IQ, Jones GH, Bunch ST (1987) Behavioural, biochemical and histochemical effects of different neurotoxic amino acids injected into into nucleus basalis magnocellularis of rats. Neuroscience 20:653–669

Dunnett SB, Evenden JL, Iversen SD (1988) Delay-dependent short-term memory impairments in aged rats. Psychopharmacology (Berlin) (in press)

Dunnett SB, Badman F, Rogers DC, Evenden JL, Iversen SD (1989) Cholinergic grafts in the neocortex or hippocampus of aged rats: reduction of delay-dependent deficits in the delayed non-matching to position task. Exp Neurol (in press)

Dunnett SB Rogers DC, Jones GH (1989) Effects of nucleus basalis magnocellularis lesions on delayed matching and non-matching to position tasks: disruption of conditional discrimination learning but not of short-term memory. Eur J Neurosci (in press)

Etherington R, Mittleman G, Robbins TW (1987) Comparative effects of nucleus basalis and fimbria-fornix lesions on delayed matching and alternation tests of memory. Neurosci Res Commun 1:135–143

Etienne P, Robitaille Y, Wood P, Gauthier S, Nair NPV, Quirion R (1986) Nucleus basalis, neuronal loss, neuritic plaques and choline acetyltransferase activity in advanced Alzheimer's disease. Neuroscience 19:1279–1291

Everitt BJ, Robbins TW, Evenden JL, Marston HM, Jones GH, Sirkiä TE (1987) The effects of excitotoxic lesions of the substantia innominata, ventral and dorsal globus pallidus on the acquisition and retention of a conditional visual discrimination: implications for cholinergic hypotheses of learning and memory. Neuroscience 22:441–469

Fine A, Dunnett SB, Björklund A, Iversen SD (1985a) Cholinergic ventral forebrain grafts into the neocortex improve passive avoidance memory in a rat model of Alzheimer's disease. Proc Natl Acad Sci USA 82:5227–5230

Fine A, Dunnett SB, Björklund A, Clarke DJ, Iversen SD (1985b) Transplantation of embryonic ventral forebrain neurons to the neocortex of rats with lesions of nucleus basalis magnocellularis. I. Biochemical and anatomical observations. Neuroscience 16:769–786

Fisher A, Hanin I (1986) Potential animal models for senile dementia of Alzheimer's type, with an emphasis on AF64A-induced toxicity. Annu Rev Pharmacol Toxicol 26:161–181

Flicker C, Ferris SF, Bartus RT, Crook T (1984) Effects of aging and dementia upon recent visuo-spatial memory. Neurobiol Aging 5:75−83

Gage FH, Björklund A (1986) Cholinergic grafts into the hippocampal formation improve spatial learning and memory in aged rats by an atropine sensitive mechanism. J Neurosci 6:2837−2847

Gage FH, Björklund A, Stenevi U, Dunnett SB, Kelly PAT (1984a) Intrahippocampal septal grafts ameliorate learning impairments in aged rats. Science 225:533−536

Gage FH, Dunnett SB, Björklund A (1984b) Spatial learning and motor deficits in aged rats. Neurobiol Aging 5:43−48

Gage FH, Kelly PAT, Björklund A (1984c) Regional changes in brain glucose metabolism reflect cognitive impairments in aged rats. J Neurosci 4:2856−2866

Glick SD, Jarvik ME (1970) Differential effects of amphetamine and scopolamine on matching performance of monkeys with lateral frontal lesions. J Comp Physiol Psychol 73:307−313

Hagan JJ, Morris RGM (1988) The cholinergic hypothesis of memory: a review of animal experiments. In: Iversen LL, Iversen SD, Snyder SY (eds) Handbook of psychopharmacology, vol 20. Plenum, New York, pp 237−323

Hardy JA, Mann DMA, Wester P, Winblad B (1986) An integrative hypothesis concerning the pathogenesis and progression of Alzheimer's disease. Neurobiol Aging 7:489−502

Heise GA (1975) Discrete trial analysis of drug action. Fed Proc 34:1898−1903

Hodos W (1970) Evolutionary interpretation of neural and behavioral studies of living vertebrates. In: Schmitt FO (ed) The neurosciences IInd study program. Rockefeller University Press, New York, pp 26−39

Höhmann C, Antuono P, Coyle JT (1988) Basal forebrain cholinergic neurons and Alzheimer's disease. In: Iversen LL, Iversen SD, Snyder SY (eds) Handbook of psychopharmacology, vol 20. Plenum, New York, pp 69−105

Kesner RP, Bierley RA, Pebbles P (1981) Short-term memory: the role of d-amphetamine. Pharmacol Biochem Behav 15:673−676

Kubanis P, Zornetzer SF (1981) Age-related behavioral and neurobiological changes: a review with an emphasis on memory. Behav Neural Biol 31:115−172

Luiten PGM, Gaykema RPA, Traber J, Spencer DG (1987) Cortical projection patterns of magnocellular basal nucleus subdivisions as revealed by anterogradely transported *phaseolus vulgaris* leucoagglutinin. Brain Res 413:229−250

Mann DMA (1988) Neuropathological and neurochemical aspects of Alzheimer's disease. In: Iversen LL, Iversen SD, Snyder SY (eds) Handbook of psychopharmacology, vol 20. Plenum, New York, pp 1−67

Mishkin M (1978) Memory in monkeys severely impaired by combined but not separate removal of amygdala and hippocampus. Nature 273:297−298

Morris RG, Kopelman MD (1986) The memory deficits in Alzheimer-type dementia: a review. Q J Exp Psychol [A] 38:575−602

Nilsson OG, Shapiro ML, Gage FH, Olton DS, Björklund A (1987) Spatial learning and memory following fimbria-fornix transection and grafting of fetal septal neurons to the hippocampus. Exp Brain Res 67:195−215

Olton DS, Walter JA, Gage FH (1978) Hippocampal connections and spatial discrimination. Brain Res 139:295−308

Pearson RCA, Powell TPS (1987) Anterograde vs retrograde degeneration of the nucleus basalis medialis in Alzheimer's disease. In: Wurtman RJ, Corkin SH, Growden JH (eds) Alzheimer's disease: Advances in basic research and therapies. Centre for Brain Sciences and Metabolism Trust, Cambridge/MA, pp 123−133

Pearson RCA, Sofroniew MV, Cuello AC, Powell TPS, Eckenstein F, Esiri MM, Wilcock GK (1983) Persistence of cholinergic neurons in the basal nucleus in a brain with senile dementia of the Alzheimer's type demonstrated with immunohistochemical staining for choline acetyltransferase. Brain Res 289:375−379

Pearson RCA, Esiri MM, Hiorns RW, Wilcock GK, Powell TPS (1985) Anatomical correlates of the distribution of the pathological changes in neocortex in Alzheimer's disease. Proc Natl Acad Sci USA 82:4531−4534

Perry EK, Perry RH, Blessed G, Tomlinson BE (1977) Necropsy evidence of central cholinergic deficits in senile dementia. Lancet I:189

Perry EK, Tomlinson BE, Blessed G, Bergmann K, Gibson PH, Perry RH (1978) Correlation of cholinergic abnormalities with senile plaques and mental test scores in senile dementia. Br Med J 2:1457–1459

Price DL (1986) New perspectives on Alzheimer's disease. Annu Rev Neurosci 9:489–512

Ridley RM, Murray TK, Johnson JA, Baker HF (1986) Learning impairment following lesion of the basal nucleus of Meynert in the marmoset: modification by cholinergic drugs. Brain Res 376:108–116

Rothblatt LA, Hayes LL (1987) Short-term object recognition memory in the rat: nonmatching with trial-unique junk objects. Behav Neurosci 101:587–590

Sahakian BJ, Morris EG, Evenden JL, Heald A, Levy R, Philpot M, Robbins TW (1988) A comparative study of visuospatial memory and learning in Alzheimer-type dementia and Parkinson's disease. Brain 111:695–718

Sjögren J (1952) Clinical aspects of morbus Alzheimer and morbus Pick. Acta Psychiatr Neurol Scand [Suppl] 82:69–115

Smith G (1988) Animal models of Alzheimer's disease: experimental cholinergic denervation. Brain Res Rev 13:103–118

Wallace J, Steinert PA, Scobie SR, Spear NE (1980) Stimulus modality and short-term memory in rats. Anim Learn Behav 8:10–16

Warburton DM (1974) The effects of scopolamine on a two-cue discrimination. Q J Exp Psychol 26:395–404

Wenk GL, Olton DS (1987) Basal forebrain cholinergic neurons and Alzheimer's disease. In: Coyle JT (ed) Animal models of dementia: a synaptic neurochemical perspective. Liss, New York, pp 81–101

Whitehouse PJ, Price DL, Struble RG, Clark AW, Coyle JT, DeLong MR (1982) Alzheimer's disease and senile dementia: loss of neurons in the basal forebrain. Science 215:1237–1239

Wilcock GK, Esiri MM, Bowen DM, Smith CCT (1982) Alzheimer's disease: correlation of cortical choline acetyltransferase activity with the severity of dementia and histological abnormalities. J Neurol Sci 57:407–417

Zola-Morgan S, Squire LR, Mishkin M (1982) The neuroanatomy of amnesia: amygdala-hippocampus vs temporal stem. Science 218:1337–1339

Pathophysiology of the Subcortically Deafferented Hippocampus: Improvement and Deterioration of Function by Fetal Grafts

G. Buzsáki, and *F. H. Gage*

Summary

In Alzheimer's disease multiple abnormalities of cortical cells and subcortical neuronal groups occur, including major losses of basal cholinergic neurons and serotonergic and noradrenergic cells. In a series of studies in the rat we investigated the physiological consequences of subcortical deafferentation of the hippocampus. We found that in the absence of cholinergic and aminergic inputs, long-term potentiation in the dentate gyrus is severely impaired, and epileptic interictal discharges with concurrent neuronal population bursts will develop after the deafferentation. The behavioral consequences of subcortical denervation are as detrimental as removing the whole hippocampus. We suggest that the memory deficit in Alzheimer's disease can be explained by "deafferentation" due to damage of the subcortical nuclei.

We attempted to reverse the pathophysiological changes by grafting locus coeruleus, hippocampus premordia, and histamine-rich cells of the supramammillary region of the fetal brain. With transplants containing histamine neurons we could restore long-term potentiation in the dentate gyrus. Grafts of the locus coeruleus substantially decreased the incidence of interictal spikes and protected the animals from picrotoxin-induced behavioral seizures. Hippocampal grafts, on the other hand, substantially disrupted hippocampal function. In rats with hippocampal grafts the incidence of interictal spikes was the highest of all the groups, and half of the rats displayed occasional behavioral seizures. Recording from the graft itself revealed that the hippocampal graft possessed epileptic activity, and anatomical and physiological connections between the graft and host were a prerequisite for propagation of epileptic seizures to the host. These adverse effects of the transplanted hippocampus should draw attention to the need for chronic and detailed physiological studies of the grafted neuronal tissue before possible application of the transplantation method in clinical settings.

Introduction

The hippocampal system has been strongly implicated in learning and memory. Damage to it produces an amnesic syndrome in people (Milner et al. 1968), monkeys (Mishkin 1982), and rats (O'Keefe and Nadel 1978; Olton et al. 1979; Isaacson and Pribram 1984). One of the most effective lesions in rats may be produced by transecting the fimbria-fornix complex. Indeed, damage to the fimbria-fornix results in as serious a behavioral deficit as removing the whole hippocampus (O'Keefe and Nadel 1978; Olton et al. 1979; Jarrard 1986). These findings indicate that interrupting the major communication fiber systems between the hippocampus and the basal forebrain, hypothalamus, and brainstem completely destroys the ability of the hippocampal neuronal network to carry out the required processing of neocortically mediated information.

In Alzheimer's disease (AD) multiple abnormalities of cortical cells and subcortical neuronal groups have been reported, including major losses of basal forebrain cholinergic neurons and serotonergic and noradrenergic cells (Terry et al. 1981; Whitehouse et al. 1982; Martin and Barchas 1986). Diffuse damage of the neocortex and hippocampus contributes significantly to the disorder. Based on animal experiments, we believe that the loss of subcortically-mediated neurotransmitters alone may explain several of the symptoms (Buzsáki et al. 1988a; Steriade and Buzsáki 1989; Vanderwolf 1988). For example, the profile of the behavioral impairment following circumscribed damage to the septal area is in many ways similar to the consequences of hippocampal damage (Winson 1978).

Why is the hippocampus so dependent on the integrity of subcortical inputs, which, in conventional psychological terms, convey information about the motivational, emotional, arousal, and autonomic states of the animal?

In this chapter we summarize our physiological experiments with the chronically denervated hippocampus and indicate that isolation of the hippocampus from its subcortically-mediated neurotransmitters results in epileptiform activity and an impaired long-term potentiation (LTP) mechanism (Buzsáki and Gage 1989; Buzsáki et al. 1988b). Furthermore, we describe our findings with neuronal grafts that were designed to reverse the lesion-induced impairments and point out both beneficial and adverse effects of fetal tissue transplants. Our guiding hypothesis is that the lack of synaptic plasticity (LTP) and/or the abnormal spontaneous electrical patterns in the subcortically denervated hippocampus may be the underlying neuronal mechanisms of the behavioral impairment. The method of neuronal transplantation allowed us to test this hypothesis by grafting cell groups of the fetal brain that normally innervate the hippocampus.

Connections of the Hippocampus

Electrophysiological and anatomical research indicates that the hippocampal formation contains a unidirectional neuronal circuit consisting of the following cellular areas and pathways: entorhinal cortex (perforant path) — dentate gyrus (mossy fibers) — CA3 region (cytoarchitectural field three; Schaffer collaterals) — CA1 region (alveus) — subiculum — entorhinal cortex (Andersen et al. 1971; Finch et al. 1983). The two hippocampi are interconnected with the ventral hippocampal commissure and indirectly via the presubiculum — contralateral entorhinal cortex (psalterium dorsale) paths (Bartesaghi et al. 1988).

The major input to the hippocampus arrives from the entorhinal cortex and the fibers terminate on the dendrites of the dentate granule cells and the distal portion of the apical dendrites of the pyramidal cells of the CA1−3 regions (Steward

1976). The terminals of both the perforant path and commissural inputs release excitatory amino acids (Fagg 1985), and the two pathways together provide approximately 95% of the extrinsic afferents to the hippocampus (Buzsáki 1984).

The remaining afferents derive from subcortical nuclei. These fibers (only 5% of all afferents) must control the hippocampal circuitry in a very unique way, since damage of a comparable percentage of fibers of other inputs has only minor behavioral consequences (Olton et al. 1979). The subcortical projections to the hippocampus involve cholinergic and gamma-aminobutyric acid-(GABA)ergic fibers from the septal area (Amaral and Kurz 1985), noradrenergic fibers from the locus coeruleus (Lindvall and Björklund 1974), serotonergic fibers from the raphe nuclei (Steinbusch 1981), histaminergic fibers from the hypothalamus (Wouterlood et al. 1986), and several minor pathways from over 20 other subcortical nuclei (Wyss et al. 1979). In the rat the subcortical afferents to the hippocampus project via the fimbria-fornix, the cingulum bundle and the supracallosal stria. A minor portion of afferents to the ventral tip of the hippocampus travel parallel with the amygdalofugal path (Gage et al. 1983). Although the hippocampus sends efferents to the lateral septum, hypothalamus, thalamus, and other subcortical targets, the overwhelming majority of efferent information is fed back to the entorhinal cortex (Finch et al. 1983).

To summarize, the entorhinal cortex and hippocampus are interconnected by a series of undirectional associational pathways and all cellular fields receive subcortical afferents. Physiologically, the entorhinal cortex − hippocampal formation − entorhinal cortex circuitry can be conceived as a positive feedback loop, whose amplification is normally controlled by the subcortical afferents.

Neuronal Grafting

When fetal tissue is grafted into the adult brain it develops in an environment substantially different from the developing brain. Even when placed into a homologous structure (e.g., hippocampal tissue into the hippocampus), the rate of changes in the graft and the injured host tissue are expected to be different. The transplanted tissue has to build up its own cytoarchitecture and give rise to intrinsic and extrinsic connections. The architectural development seems to be autonomously programmed since lamination and other cytoarchitectonic rules are preserved to a considerable degree even in culture dishes. For example, dentate granule cells which undergo final mitosis postnatally develop a relatively preserved lamination even when transplanted as cell suspension (our unpublished findings). Interestingly, the already existing regular cytoarchitecture, e.g., CA fields of the hippocampus, deteriorates during the course of grafting.

A second issue regarding the maturation of grafted tissue is whether the different cell types, characteristic to a given structure, develop in normal proportion and quantity. Very little is known about the factors that regulate the expression of the various neuronal enzymes for producing various neurotransmitters and modulators. Often, coexistence of transmitters and neuromodulators or coexistence of classical neurotransmitters in the same neuron is explained simply as statistical variability of the developmental process. Are these processes preprogrammed in the neuron or subject to environmental influences? If the latter is the case then the grafted tissue might provide a different neural profile to the normally developing one.

A third issue concerns the lack of equipotentiality of the various neuronal systems with regard to growth and regeneration. For example, monoaminergic and GABAergic neurons appear to sprout more extensively than axon terminals of glutaminergic neurons (Sunde et al. 1984; Aguayo 1985). Consequently, the proportions of the graft- and host-targeted afferents may vary considerably.

An essential final question to ask, of course, is how these variables influence the benevolent or potentially adverse effects of the grafted fetal tissue on the function of the normal brain. To date, very few generalizations can be made regarding the mechanisms of action of neuronal grafts in different systems, and the reader is referred to recent reviews on this topic (Björklund et al. 1987; Buzsáki and Gage 1988).

It is usually tacitly assumed that deleterious effects of neural implants are due to graft-induced scar or cyst formation (Kruger et al. 1986), ventricular enlargement, interruption of host connections, defects in blood-brain barrier (Rosenstein and Brightman 1986), and degenerative changes in the host parenchyma (Dalrymple-Alford et al. 1988). Conversely, the most successful and functionally effective improvement is expected from implants with extensive afferent and efferent graft–host connections (Björklund et al. 1987; Sotelo and Alvarado-Mallart 1985; see also other chapters in this volume).

In this chapter we summarize our experience with hippocampal grafts and will demonstrate that the grafted hippocampus serves as an epileptic focus which, in many instances, is capable of kindling the host brain to express behavioral seizures. We provide evidence that this adverse effect of the hippocampal graft is not due to anatomical damage to the host but is a genuine feature of the grafted tissue which requires physiological connections between the implant and the host: the better the integration, the higher the probability of the epileptic manifestations. These experiments draw attention to the need for chronic and detailed physiological studies of the transplanted neuronal tissue before possible application of the method in clinical settings for promoting recovery from brain damage or age-related degeneration.

The Grafting Procedure

The three main procedures used for grafting hippocampal tissue are
1. placing solid pieces of the whole fetal hippocampus into a prepared cavity under visual control (Kromer et al. 1981),
2. injection of small pieces of the dissected fetal hippocampus by means of a glass capillary (Sunde and Zimmer 1981), and
3. injection of trypsin-dissociated suspension of fetal neuronal tissue (Björklund et al. 1983). Each method has its advantages and disadvantages depending on the questions asked.

The degree of anatomical integration of the graft with the host depends on several factors involving the age of the donor, age of the recipient, the grafting procedure, pretreatment and storage of the donor tissue, and the functional state of the recipient structure. These variables and the various anatomical connections of the grafted hippocampus with the host brain have been recently reviewed (Zimmer et al. 1988).

In our experiments we used either solid grafts placed into a cavity made by interrupting the subcortical connections of the hippocampus or cell suspensions directly injected into the host hippocampus of adult rats. In other experiments we removed the hippocampus of newborn rats by aspiration and replaced it with a fetal hippocampus. The age of the donors was between embryonic days 15 and 17.

Electrical Activity in the Subcortically Denervated Hippocampus

Removal of subcortical afferents and efferents of the host hippocampus is made by aspirating the fimbria, the dorsal fornix, the ventral hippocampal commissure, the corpus callosum, and the overlying cingulate cortex. The lesion eliminates the afferent brainstem projections from the locus coeruleus, the dorsal and medial raphe nuclei, the supramammillary region, and other minor nuclei, as well as the cholinergic and GABAergic forebrain projections from the medial septal area and the nucleus of the diagonal band of Broca. Although subcortical inputs to the ventral tip of the hippocampus remain intact, the dorsal part of the hippocampus remains permanently denervated (Gage et al. 1983). The fimbria-fornix (FF) lesion also removes all subcortical efferents from the hippocampus.

In the subcortically denervated (FF-lesioned) hippocampus large amplitude (3–8 mV) and short duration (20–20 ms) EEG spikes are present several months after the operation. These large transients can be distinguished from the physiological sharp waves of the normal hippocampus based on their amplitude, duration, waveform, and behavioral correlates (Buzsáki 1986). The short duration, large amplitude transients are regarded as interictal spikes (IIS) (Buzsáki et al. 1988b).

The FF-lesioned hippocampus is strikingly more prone to seizures than the normal hippocampus. Low frequency (5 Hz, 6 s) stimulation of the perforant path reliably induced epileptic afterdischarges in both normal and lesioned rats, but significantly lower current levels were required in FF-lesioned animals.

An additional difference between normal and FF-lesioned animals was the sensitivity to the GABA-blocker, picrotoxin. Picrotoxin-induced IIS also occurred in normal rats but the incidence of IIS in FF-lesioned rats was significantly higher (Buzsáki et al. 1989b).

The robust differences between the normal and the FF-lesioned hippocampus in terms of IIS, afterdischarge threshold, propagation of afterdischarges, and sensitivity to picrotoxin, and the permanent nature of these changes make the subcortically denervated hippocampus an ideal model with which to study the effects of neuronal grafts on modulating cellular excitability. It may well be that these pathophysiological changes in the denervated hippocampus interfere not only with the function of the hippocampus itself but the epileptic activity may spread to other structures, thereby deteriorating behavioral performance further (Buzsáki and Gage 1988).

Absence of LTP in the Subcortically Denervated Dentate Gyrus

An extensively studied neurophysiological model for a cellular mechanism of memory trace formation in the mammalian brain is LTP (Bliss and Lømo 1973). Several lines of evidence suggest that the mechanism underlying the trigger event of LTP is a large depolarization of the postsynaptic membrane during the conditioning train. Large depolarization of the postsynaptic cell for a sufficient amount of time appears to relieve a Mg^{2+}-block of the N-methyl-D-aspartate (NMDA) subtype of glutamate-activated channel (Collingridge et al. 1983). Only when this channel is unblocked can it conduct in response to glutamate and allow Ca^{2+} influx. Recently, we have found that LTP is lost permanently in the dentate gyrus following FF lesion (Fig. 1) (Buzsáki and Gage 1989). We hypothesize that the loss of synaptic plasticity in the subcortically deafferented dentate gyrus would be another potential pathomechanism of the memory impairment at the behavioral level.

The lack of LTP in the deafferented dentate gyrus may be explained by the following alternative hypotheses. (a) The importance of noradrenaline, serotonin, and acetylcholine in LTP generation has been previously suggested (Robinson and Racine 1982; Bliss et al. 1983; Hopkins and Johnston 1984; Stanton and Sarvey 1985). A common denominator among the neurotransmitters of subcortical origin might be that they are known to decrease a Ca^{2+}-dependent K^+-conductance in pyramidal and granule cells (Langmoen et al. 1981; Cole and Nicoll 1984; Haas and Konnerth 1983; Haas and Rose 1987) which follows burst firing of these cells. In the absence of acetylcholine and monoamines the K^+-mediated afterhyperpolarization (AHP) is substantially enhanced. It is conceivable that due to the enhanced AHP the neuron cannot be depolarized for a sufficient amount of time in response to the conditioning train and therefore the NMDA channels fail to open. (b) As discussed above large amplitude epileptic IIS occur in the subcortically denervated hippocampus. During the IIS the highly synchronous bursts of

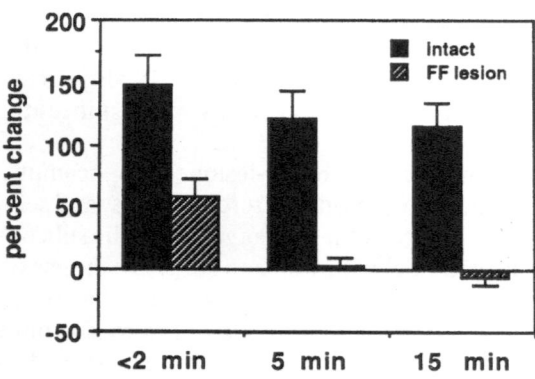

Fig. 1. Percentage change in population spike amplitude (mean+SEM) in the dentate gyrus in intact and fimbria-fornix deprived (FF) rats. (Reprinted from Buzsáki and Gage, 1989)

neurons may depolarize their targets sufficiently to produce LTP in those neurons. In the FF-deprived dentate gyrus the LTP mechanism may therefore become saturated due to the chronic IIS, and this could explain the absence of further LTP by electrical stimulation.

In a series of experiments still in progress we attempted to restore LTP in the denervated dentate gyrus by grafting various subcortical nuclei of the fetal brain. To date, we found that histamine-rich cells, derived from the mammillary region of the fetal brain, were able to restore the lost synaptic plasticity in the dentate gyrus completely, when transplantation was made at the time of the FF damage. Segal (this volume) reports that raphe grafts were also effective in improving LTP in animals with previous chemical damage of the serotonergic system.

Graft Action on the Denervated Hippocampus

Based on previous histological and tissue culture experiments and the importance of GABA in the control of epileptic activity the hippocampal graft, rich in GABAergic neurons, was a promising choice to moderate the epileptic tendencies of the denervated hippocampus. Developmental and tissue culture investigations reported that the relative proportion of GABAergic cells is higher in the hippocampus at embryonic day 15 than in the adult (Walker and Peacock 1982; Hoch and Dingledine 1986). In addition, recent anatomical studies indicated the presence and high prevalence of GABA-immunoreactive neurons in hippocampal grafts. Contrary to our expectations, however, we found in our initial electrophysiological studies that the transplanted hippocampus displays continuous epileptic spikes (Buzsáki et al. 1987a, b). In subsequent experiments we studied the effects of hippocampal grafts on the host hippocampus and compared it with the consequences of locus coeruleus transplants (Buzsáki et al. 1988b).

We have chosen noradrenergic cells for grafting because previous studies have established that:
1. The ascending noradrenergic system may have a seizure suppressant action (McIntyre and Edson 1982).
2. Intrahippocampal grafts of locus coeruleus have been shown to establish anatomical and physiologically active connections with the host cells (Björklund et al. 1979).
3. Cografting of fetal locus coeruleus and hippocampus in the anterior chamber of the eye increased the threshold of stimulation-induced epileptic discharges (Taylor et al. 1980).
4. Locus coeruleus grafts retarded the development of kindling- induced seizures in animals with previous chemical lesions of the catecholamine system (Barry et al. 1987).

Electrical Activity in the Transplanted Hippocampus

In our initial experiments we studied the neuronal patterns of the grafted hippocampus and its interaction with the host brain. The transplanted animals were

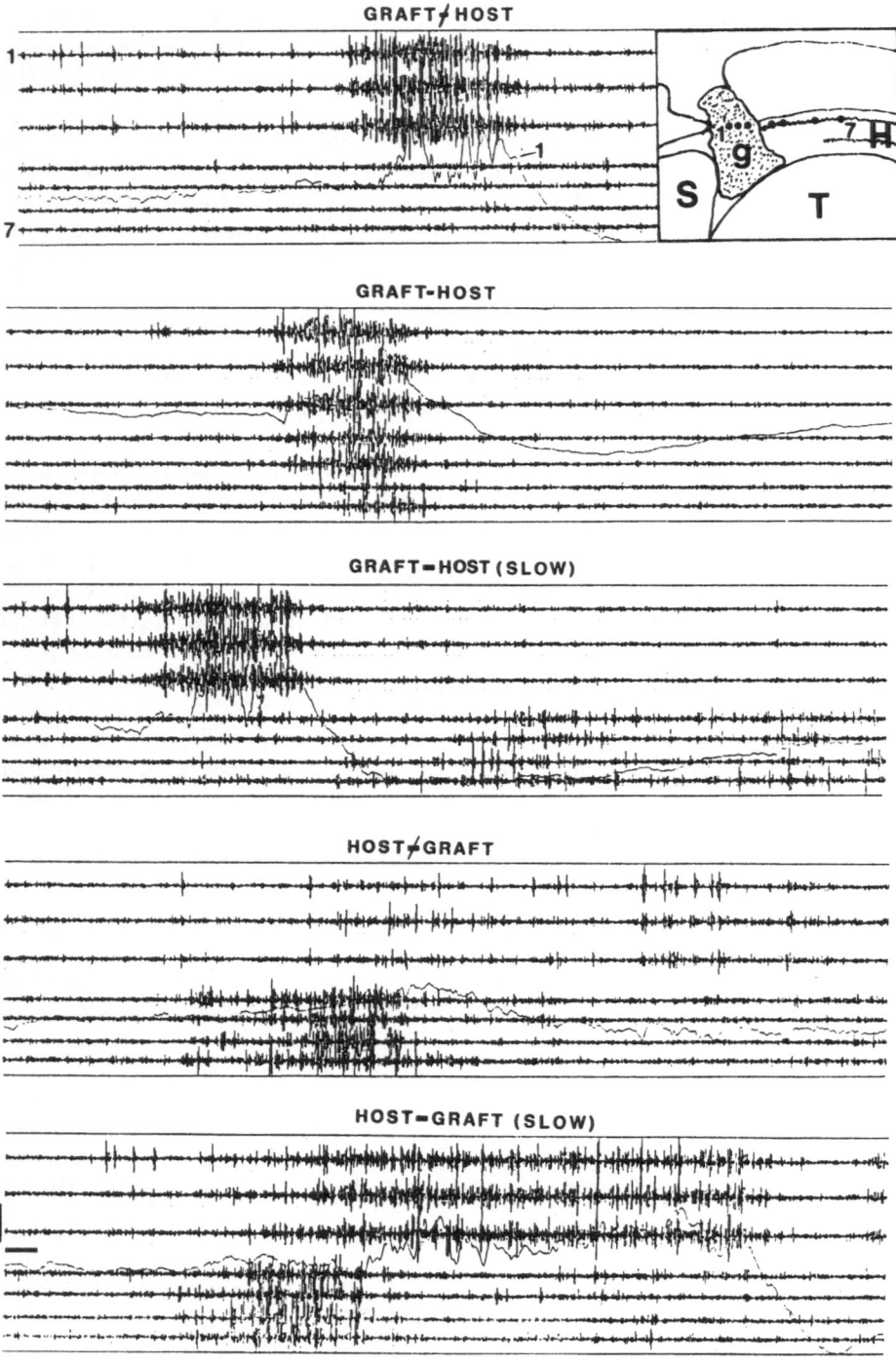

Fig. 2

equipped with recording and stimulating electrodes 5—9 months after the grafting procedure. For studying the reciprocal physiological connections between the grafted and host hippocampi a 16-wire microelectrode probe with 250 μm horizontal tip separations was used (Buzsáki et al. 1989a). All experiments were carried out in freely moving rats.

Solid hippocampal grafts displayed highly synchronous population bursts and concurrent large amplitude EEG spikes. The EEG spike and concurrent population burst could be triggered by stimulating the perforant path or the host hippocampus. Both complex-spike cells (putative pyramidal cells; Ranck 1973; Fox and Ranck 1986) and single-spike cells (interneurons or granule cells; Fox and Ranck 1986; Buzsáki et al. 1983) were observed in the graft, and both cell types fired maximally during the EEG spike. In sharp contrast to the pyramidal cells of the normal hippocampus (Fox and Ranck 1986), putative pyramidal cells in the graft often responded with complex-spike patterns to host stimulation, indicating the impairment of recurrent inhibition (Andersen et al. 1963).

The speed of propagation of the EEG spike and concurrent population burst within the graft and across the graft — host hippocampus interface was either slow (<0.5 m/s) or fast (>3 m/s) (Fig. 2). Large amplitude, short duration spikes usually propagated at a high speed, while low amplitude, wider spikes with broad population bursts spread at a lower velocity.

Occasional seizures were also seen in the graft which spread to the host hippocampus (Fig. 3; Buzsáki et al. 1987a). Afterdischarges in the host hippocampus could occur virtually simultaneously with those of the graft or could be slowly built up with a delay of several seconds. Subsequent histological examination revealed that in animals with relatively loose connections between the graft and the host the spread of EEG spikes and epileptic seizures occurred at a low probability, while in rats with efficient spread of the electrical activity the graft fused into the host hippocampus.

We also examined synaptic plasticity in the hippocampal graft. High frequency stimulation of either the host hippocampus or direct stimulation of the graft itself induced robust facilitation of the evoked cellular discharges and field potentials to subsequent single pulses. A major difference between the LTP (Bliss and Lømo 1973) of the synaptic efficacy in the graft and in the host hippocampus was that significantly lower currents were required to induce synaptic enhancement in the graft than in the intact host. This finding is a further indication of the impaired inhibition in the transplanted hippocampus, since blockade of GABA-ergic inhibition is known to facilitate the induction of LTP in the hippocampus (Wigstrom and Gustafsson 1983). Besides the enhancement of evoked responses

---

Fig. 2. Simultaneous recording of electrical activity from a solid hippocampal graft *(g)* placed in the fimbria-fornix cavity and from the host hippocampus *(H)* in the freely moving rat. An array of 16 microelectrodes with 250 μm horizontal tip separations was implanted. The electrode assembly can be moved vertically by means of three screws. A 16-channel MOSFET-input preamplifier integrated in the headstage is used to eliminate movement artefacts. *Circles* in the inset indicate the tips of the electrodes used for the recordings shown. Also shown is a filtered record of an interictal spike recorded simultaneously from electrode 1. Note bidirectional spread or lack of spread of the interictal bursts. Calibrations: 2 mV (slow wave), 200 μV (unit); 10 ms. *S*, septum; *T*, thalamus. (Reprinted from Buzsáki et al. 1989c)

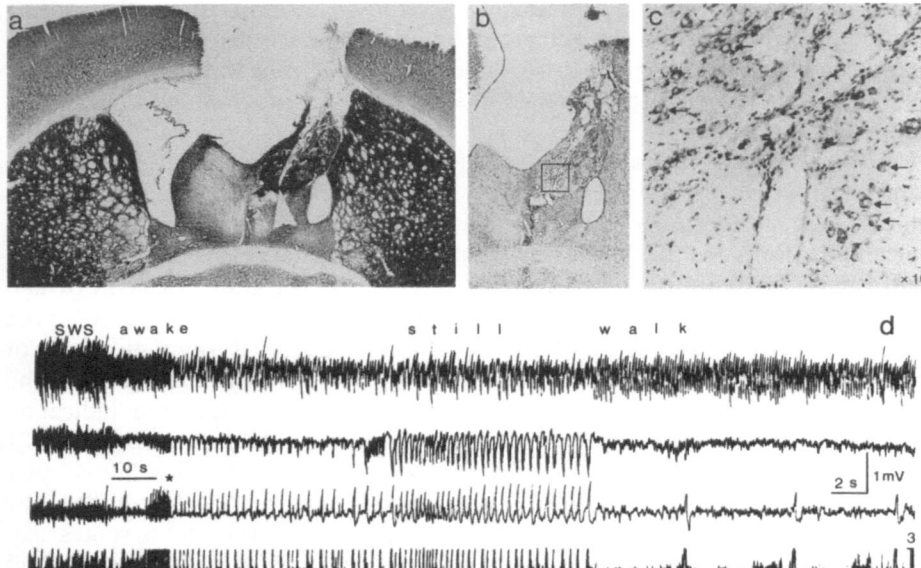

Fig. 3a–d. Propagation of electrical activity from graft to host. **a** AChE-stained section of hippocampal transplant at the level of the electrode penetration *(white arrowhead).* **b** Nissl-stained section of the same region. The tip area *(box)* is shown at a higher magnification in **c** *Arrows* in **c** indicate pyramidal cells. **d** Spontaneous seizure recorded during slow wave sleep *(SWS)* – awake transition. Note spread from graft to host. *First trace,* contralateral (intact) hippocampus; *second trace,* ipsilateral host hippocampus; *third* trace, EEG from the graft; *fourth trace,* discriminated multiple unit activity. *Asterisk* indicates change in recording speed. (From Buzsáki et al. 1987a)

we found that high-frequency stimulation substantially increased both the incidence and amplitude of the population cellular bursts and concurrent EEG spikes. This robust increase of the spontaneous spiking remained above baseline levels for several hours. The presence of LTP mechanisms in the grafted hippocampus suggests either that the graft has become reinnervated by the necessary subcortical fibers or that LTP in the CA fields does not necessarily require the presence of subcortical inputs. Due to the mixed nature of neurons in the hippocampal graft we could not establish whether LTP was mediated by granule cells or pyramidal neurons.

In agreement with previous anatomical studies (Frotscher and Zimmer 1987; Robain et al. 1987; Freund and Buzsáki 1988) numerous large GABA- and parvalbumin-immunoreactive neurons were present in the hippocampal grafts. This was in contrast with the host hippocampus where the numbers of parvalbumin-immunoreactive neurons were severely reduced, especially in the CA3 area adjacent to the graft (Fig. 4).

Subsequent electron microscopic examination of the hippocampal grafts revealed the absence of a strategically important inhibitory neuronal type, the

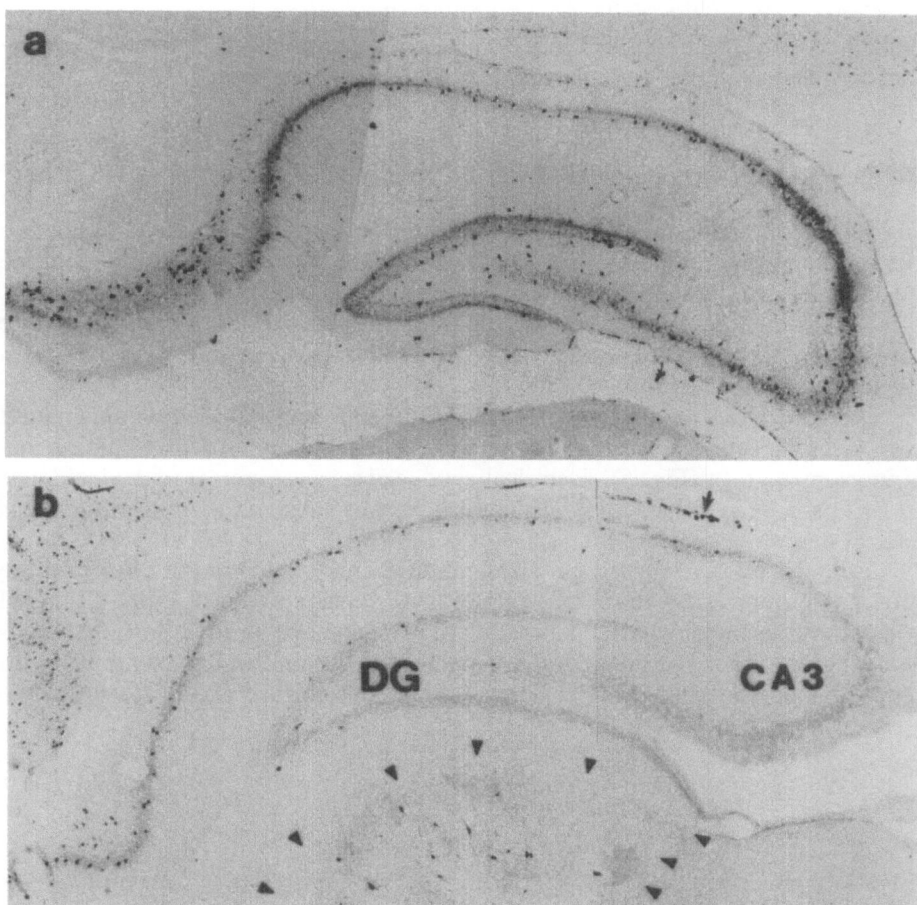

Fig. 4a, b. Severe reduction of parvalbumin-immunoreactive neurons in the host hippocampus with hippocampal graft. **a** Normal control. **b** Fimbria-fornix lesioned hippocampus with hippocampal graft. The border of the transplanted hippocampus is marked by *arrowheads*. Note virtual absence of immunoreactive neurons in the hilus and *CA3* region and numerous, large parvalbumin-positive neurons in the graft. *Arrow*, blood cells; *DG*, dentate gyrus

chandelier or axoaxonic cell (Freund and Buzsáki 1988). In addition, several types of asymmetric, non-GABAergic synapses were found on the somata of pyramidal cells. Such presumably excitatory synapses are not present on the cell bodies of pyramidal neurons in the normal hippocampus (Kosaka 1983; Seress and Ribak 1985).

The excessive number of collaterals may be explained by assuming that the developing axons in the graft terminate preferentially on the dendrites and somata of neighboring neurons since fibers ingrowing from the host are limited and not competing for the available postsynaptic sites.

Consequently, the epileptic nature of the grafted hippocampus may be explained by the reduction of GABAergic inhibition at a specific site of the pyramidal cells and excessive mutual excitation of these neurons.

Modulation of Hippocampal Excitability by Fetal Neuronal Grafts

In order to obtain better integration with the host brain and avoid the reinnervation of the grafted tissue by subcortical afferents, in subsequent experiments we injected cell suspensions, prepared from dissociated fetal hippocampal neurons, directly into the denervated hippocampus. In the same series of experiments we investigated the effects of noradrenaline-rich cell suspension grafts prepared from the locus coeruleus region.

The occurrence of large EEG spikes was 10–30 times less frequent in animals with locus coeruleus grafts than in rats with FF lesion only, or with FF lesion and hippocampal grafts (Fig. 5). Electrically induced seizures increased the incidence of EEG spikes in all groups, but the increase was substantially less in the locus coeruleus graft group.

Administration of picrotoxin also had differential behavioural effects on the different groups. Less than 15% of rats with locus coeruleus grafts displayed behavioural seizures, while all animals with FF lesion only or FF lesion and hippocampal grafts showed behavioural seizures within 30 min. In half of the rats with hippocampal transplants at least one spontaneous seizure was observed in the

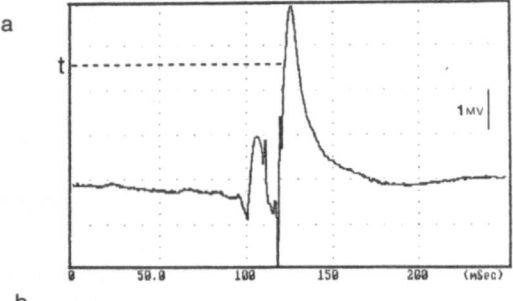

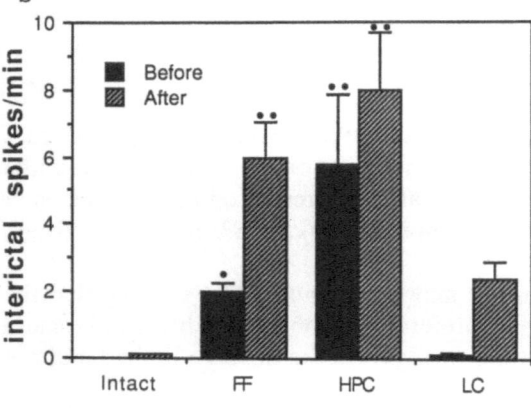

Fig. 5a, b. Facilitation and suppression of interictal spikes in the hippocampus by fetal neuronal tissue. **a** A representative interictal spike recorded in a fimbria-fornix lesioned rat. **b** Frequency of interictal spikes before *(black columns)* and one day after *(striped columns)* six hippocampal seizures (one seizure/day) induced by electrical stimulation of the perforant path. *Vertical bars,* standard errors of the means. *Dots* indicate significant differences from both intact and LC groups. (From Buzsáki et al. 1988b Proc Natl Acad Sci USA)

home cage prior to electrode implantation. None of the other animals was observed to produce spontaneous behavioural seizures. A further difference between the groups was reflected by the behavioural consequences of perforant path stimulation. Stimulation in 15% of the hemispheres in the FF group induced behavioural seizures, including rearing and falling (Racine 1972), during the very first hippocampal afterdischarges. In the group with hippocampal grafts this figure was 50%, while perforant path stimulation-induced hippocampal afterdischarges were not followed by behavioral manifestations in animals with locus coeruleus grafts and intact controls (Buzsáki et al. 1988b), except "wet-dog" shakes and grooming.

The behavioural consequences of the epileptic patterns in the hippocampal grafts may be explained by assuming that, as with the epileptic afterdischarges induced by daily electrical stimulation of various regions of the limbic system (Goddard et al. 1969; Racine 1972), the graft served as a seizure focus which eventually "kindled" the host brain. We suggest that the spontaneously occurring epileptic afterdischarges in the hippocampal graft occasionally spread to the host via its functionally active anatomical connections with the host brain and repeated occurrence of these events leads to the invasion of extralimbic motor structures. Behavioral seizures triggered by stimulating the perforant path for the first time are interpreted in a similar way, namely that the spontaneously occurring seizures had already kindled extralimbic structures and stimulation thus activated the already plastic pathways. These findings also provide evidence that spread of epileptic activity from the hippocampus involves retrohippocampal structures, and the intactness of subcortical connections of the hippocampus is not essential.

Conclusions and Implications for Possible Grafting Attempts in Alzheimer's Disease

These studies indicate that neurotransmitters mediated by subcortical afferents provide electrical stability to the hippocampus. Isolation of the hippocampus from these control systems results in permanent epileptic tendencies as reflected by very large EEG spikes and significantly lowered threshold for induced seizures. In the absence of subcortically mediated neurotransmitters long-term plasticity in the dentate gyrus is impaired. Either of these pathomechanisms alone or their combination may provide a physiological explanation of the malfunctioning of the subcortically deafferented hippocampus. Similar mechanisms may be responsible for the subcortical dementia in AD.

The grafted hippocampus sustains IISs and epileptic seizures despite the presence of numerous GABAergic cells in the graft (Buzsáki et al. 1987a, b; Frotscher and Zimmer 1987; Robain et al. 1987; Zimmer 1988; Freund and Buzsáki 1988). The epileptic activity of the graft can spread to the host hippocampus via its physiologically active anatomical connections. The repetitively recurring electrical seizures in the host hippocampus may kindle extralimbic structures and result in behavioral convulsions. Conversely, cell suspension grafts of the locus coeruleus area are capable of reversing several of the epileptic symptoms.

An important implication of these findings is that it is difficult, if not impossible, to extrapolate physiological function by studying anatomical connections and neuronal distributions only. Epileptic spiking in the deafferented host hippocampus correlated with a decrease in the number of parvalbumin-immunoreactive neurons, while the incidence of these cells was high in the epileptic hippocampal graft. A further message that emerges from these experiments is that demonstration of synaptic transmission and physiological interaction does not necessarily predict improved function. In our experiments with hippocampal grafts, physiologically active synapses were a prerequisite for spreading of epileptic discharges.

Overall, while the grafting technique is an invaluable experimental tool to answer questions not addressed by other methodologies, careful and detailed physiological investigations in chronically prepared animals are of utmost importance in any clinical attempt to use fetal grafts for improving deteriorated function.

Acknowledgments. The research reviewed here was supported by grants from the J. D. French Foundation, The Sandoz Foundation for Gerontological Research, California State Department of Health, NIA and the Office of the Navy. We than Sheryl Christenson for typing the manuscript.

References

Aguayo A (1985) Axonal regeneration from injured neurons in the adult mammalian central nervous system. In: Cotman CW (ed) Synaptic plasticity. Guilford, New York, pp 457–484

Amaral DG, Kurz J (1985) An analysis of the origins of the cholinergic and non-cholinergic septal projections to the hippocampal formation of the rat. J Comp Neurol 240:37–59

Andersen P, Eccles JC, Loyning Y (1963) Recurrent inhibition in the hippocampus with identification of the inhibitory cell and its synapse. Nature 198:540–542

Andersen P, Bliss TVP, Skrede KK (1971) Lamellar organization of hippocampal excitatory pathways. Exp Brain Res 13:222–238

Barry DI, Kikvadze I, Brundin P, Bolwig TM, Björklund A, Lindvall O (1987) Grafted noradrenergic neurons suppress seizure development in kindling-induced epilepsy. Proc Natl Acad Sci USA 84:8712–8715

Bartesaghi R, Gessi T, Sperti L (1988) Electrophysiological analysis of the dorsal hippocampal commissure projections to the entorhinal area. Neuroscience 26:55–67

Björklund A, Segal M, Stenevi U (1979) Functional reinnervation of rat hippocampus by locus coeruleus implants. Brain Res 170:409–426

Björklund A, Stenevi U, Schmidt RH, Dunnett SB, Gage FH (1983) Intracerebral grafting of neuronal cell suspensions. I. Introduction and general methods of preparation. Acta Physiol Scand [Suppl] 522:1–7

Björklund A, Lindvall O, Isacson O, Brundin P, Wictorin K, Strecker RE, Clark DJ, Dunnett SB (1987) Mechanisms of action of intracerebral neural implants: studies on nigral and striatal grafts to the lesioned striatum. Trends Neurosci 10:509–516

Bliss TVP, Lømo T (1973) Long-lasting potentiation of synaptic transmission in the dentate area of the anaesthetized rabbit following stimulation of the perforant path. J Physiol (Lond) 232:331–356

Bliss TVP, Goddard GV, Riives M (1983) Reduction of long-term potentiation in the dentate gyrus of the rat following selective depletion of monoamines. J Physiol (Lond) 334:475–591

Buzsáki G (1984) Feed-forward inhibition in the hippocampal formation. Prog Neurobiol 22:131–153

Buzsáki G (1986) Hippocampal sharp waves: Their origin and significance. Brain Res 398:242–252

Buzsáki G, Gage FH (1988) Mechanisms of action of neural grafts in the limbic system. Can J Neurol Sci 15:99–105

Buzsáki G, Gage FH (1989) Absence of long-term potentiation in the subcortically denervated dentate gyrus. Brain Res (in press)

Buzsáki G, Leung LS, Vanderwolf CH (1983) Cellular bases of hippocampal EEG in the behaving rat. Brain Res Rev 6:139–171

Buzsáki G, Gage FH, Kellenyi L, Björklund A (1987a) Behavior-dependence of electrical activity of intracerebrally transplanted fetal hippocampus. Brain Res 400:321–333

Buzsáki G, Czopf J, Kondakor I, Björklund A, Gage FH (1987b) Cellular activity of intracerebrally transplanted hippocampus during behavior. Neuroscience 22:871–883

Buzsáki G, Bickford RG, Ponomareff G, Thal LJ, Mandell R, Gage FH (1988a) Nucleus basalis and thalamic control of neocortical EEG in the freely behaving rat. J Neurosci 8:4007–4026

Buzsáki G, Ponomareff GL, Bayardo F, Shaw T, Gage FH (1988b) Suppression and induction of epileptic activity by neuronal grafts. Proc Natl Acad Sci USA 85:9327–9330

Buzsáki G, Bickford RG, Ryan LJ, Young S, Prohaska O, Mandell RJ, Gage FH (1989a) Multisite recording of unit activity and field potentials in freely moving rats. J Neurosci Meth (in press)

Buzsáki G, Ponomareff GL, Bayardo F, Ruiz R, Gage FH (1989b) Neuronal activity in the subcortically denervated hippocampus: A chronic model for epilepsy. Neuroscience 28:527–538

Buszáki G, Bayardo F, Miles R, Wong RKS, Gage FH (1989c) The grafted hippocampus: an epileptic focus. Exp Neurol (in press)

Cole AE, Nicoll RA (1984) Characterization of a slow cholinergic postsynaptic potential recorded in vitro from rat hippocampal pyramidal cells. J Physiol (Lond) 352:173–188

Collingridge GL, Kehl SJ, McLennan H (1983) Excitatory amino acids in synaptic transmission in the Schaffer collateral-commissural pathway of the rat hippocampus. J Physiol (Lond) 334:33–46

Dalrymple-Alford JC, Kelche C, Cassel JC, Toniolo G, Pallge V, Will BE (1988) Behavioral deficit after intrahippocampal septal grafts in rats with selective fimbria-fornix lesions. Exp Brain Res 69:545–558

Fagg GE, Foster AC (1983) Amino acid neurotransmitters and their pathways in the mammalian central nervous system. Neuroscience 9:701–719

Finch DM, Nowlin NL, Babb TL (1983) Demonstration of axonal projection of neurons in the rat hippocampus and subiculum by intracellular injection of HRP. Brain Res 271:201–216

Fox SE, Ranck JB Jr (1986) Hippocampal theta rhythm and the firing of neurons in walking and urethane anesthetized rats. Exp Brain Res 62:495–508

Freund TF, Buzsáki G (1988) Alterations in excitatory and GABAergic inhibitory connections in hippocampal transplants. Neuroscience 27:373–386

Frotscher M, Zimmer J (1987) GABAergic nonpyramidal neurons in intracerebral transplants of the rat hippocampus and fascia dentata: a combined light and electron microscopic and immunocytochemical study. J Comp Neurol 259:266–276

Gage FH, Björklund A, Stenevi U (1983) Reinnervation of the partially deafferented hippocampus by compensatory collateral sprouting from spared cholinergic and noradrenergic afferents. Brain Res 268:27–39

Goddard GV, McIntyre DC, Leech CK (1969) A permanent change in brain function resulting from daily electrical stimulation. Exp Neurol 25:295–330

Haas HL, Konnerth A (1983) Histamine and noradrenaline decrease calcium-activated potassium conductance in hippocampal pyramidal cells. Nature 302:432–434

Haas HL, Rose GM (1987) Noradrenaline blocks potassium conductance in rat dentate granule cells in vitro. Neurosci Lett 78:171–174

Hoch DB, Dingledine R (1986) GABAergic neurons in rat hippocampal culture. Dev Brain Res 25:53–64

Hopkins WF, Johnston D (1984) Frequency-dependent noradrenergic modulation of long-term potentiation in the hippocampus. Science 226:350–352

Isaacson RL, Pribram KH (eds) (1986) The hippocampus, vols 3–4, Plenum, New York

Jarrard LE (1986) Selective hippocampal lesions and behavior: Implications for current research and theorizing. In: Isaacson RL, Pribram KH (eds) The hippocampus, vol 4. Plenum, New York

Kosaka T (1983) Axon initial segments of granule cells in the dentate gyrus: Synaptic contacts on bundles of axon initial segments. Brain Res 274:129–134

Kromer LF, Björklund A, Stenevi U (1981) Innervation of embryonic implants by regenerating axons of cholinergic septal neurons in the adult rat. Brain Res 210:153–171

Kruger S, Sievers J, Hansen C, Sadler M, Berry M (1986) Three morphologically distinct types of interface develop between adult host and fetal brain transplants: Implications for scar formation in the adult central nervous system. J Comp Neurol 249:103–116

Langmoen IA, Segal M, Andersen P (1981) Mechanisms of norepinephrine actions on hippocampal pyramidal cells in vitro. Brain Res 208:349–362

Lindvall O, Björklund A (1974) The organization of the ascending catecholamine neuron systems in the rat brain. Acta Physiol Scand 412:1–48

Martin JB, Barchas J (1986) Neuropeptides in neurologic and psychiatric disease. Raven, New York

McIntyre DC, Edson N (1982) Effect of norepinephrine depletion on dorsal hippocampal kindling in rats. Exp Neurol 77:700–704

Milner B, Corkin S, Teuber H-L (1968) Further analysis of the hippocampal amnesic syndrome: 14-year follow-up study of H.M. Neuropsychology 6:215–234

Mishkin M (1982) A memory system in the monkey. Philos Trans R Soc Lond 298:85–95

O'Keefe J, Nadel L (1978) The hippocampus as a cognitive map. Clarendon, Oxford

Olton DS, Becker JT, Handelman GE (1979) Hippocampus, space and memory. Behav Brain Sci 2:313–365

Racine RJ (1972) Modification of seizure activity by electrical stimulation. II. Motor seizure. Electroencephalogr Clin Neurophysiol 32:281–294

Ranck JB Jr (1973) Studies on single neurons in dorsal hippocampal formation and septum in unrestrained rats. I. Behavioral correlates and firing repertoires. Exp Neurol 42:461–531

Robain O, Barbin G, Ben-Ari Y, Rozenberg F, Prochiantz A (1987) GABAergic neurons of the hippocampus: development in homotopic grafts and in dissociated cell cultures. Neuroscience 23:73–86

Robinson GB, Racine RJ (1982) Heterosynaptic interactions between septal and entorhinal inputs to the dentate gyrus: long-term potentiation effects. Brain Res 249:162–166

Rosenstein JM, Brightman MW (1986) Alterations of the blood-brain barrier after transplantation of autonomic ganglia into the mammalian central nervous system. J Comp Neurol 250:339–351

Seress L, Ribak CE (1985) A substantial number of asymmetric axosomatic synapses is a characteristic of the granule cell of the hippocampal dentate gyrus. Neurosci Lett 56:21–26

Sotelo C, Alvarado-Mallart RM (1985) Cerebellar transplants: Immunocytochemical study of the specificity of Purkinje cells inputs and outputs. In: Björklund A, Stenevi U (eds) Neural grafting in the mammalian CNS. Elsevier, Amsterdam, pp 205–215

Stanton PK, Sarvey JM (1985) Depletion of norepinephrine, but not serotonin, reduces long-term potentiation in the dentate gyrus of rat hippocampal slices. J Neurosci 5:2169–2176

Steinbusch H (1981) Distribution of serotonin immunoreactivity in the central nervous system of the rat: Cell bodies and terminals. Neuroscience 6:557–618

Steriade M, Buzsáki G (1989) Parallel activation of thalamic and cortical neurons by brainstem and basal forebrain cholinergic systems. In: Steriade M, Biesold D (eds) Brain cholinergic systems. Oxford University Press, Oxford

Steward O (1976) Topographic organization of the projections from the entorhinal area to the hippocampal formation of the rat. J Comp Neurol 167:285–314

Sunde N, Zimmer J (1981) Transplantation of central nervous tissue. An introduction with results and implications. Acta Physiol Scand 63:323–335

Sunde N, Laurberg S, Zimmer J (1984) Brain grafts can restore irradiation-damaged neuronal connections in newborn rats. Nature 310:51–53

Taylor D, Seiger A, Freedman R, Olson L, Hoffer BJ (1980) Conditions for adrenergic hyperinnervation in hippocampus. II. Electrophysiological evidence from intraocular double grafts. Exp Brain Res 39:289–299

Terry RD, Peck A, deTeresa R, Schechter R, Horoupian DS (1981) Some morphometric aspects of the brain in senile dementia of the Alzheimer type. Ann Neurol 10:184–192

Vanderwolf CH (1988) Cerebral activity and behavior: Control by central cholinergic and serotonergic systems. Int Rev Neurobiol 30:225–340

Walker CR, Peacock JH (1982) Development of GABAergic function of dissociated hippocampal cultures from fetal mice. Dev Brain Res 2:541–555

Whitehouse PJ, Price DI, Struble RG, Clark AW, Coyle JT, DeLong MR (1982) Alzheimer's disease and senile dementia: Loss of neurons in the basal forebrain. Science 215:1237–1239

Wigstrom H, Gustafsson B (1983) Facilitated induction of hippocampal long-lasting potentiation during blockade of inhibition. Nature 301:603–604

Winson J (1978) Spatial deficit after septal lesion. Science 201:160–163

Wouterlood FG, Sauren YMHF, Steinbusch HVM (1986) Histoaminergic neurons in the rat brain: Correlative immunocytochemistry, Golgi impregnation, and electron microscopy. J Comp Neurol 252:227–244

Wyss JM, Swanson L, Cowan WM (1979) A study of subcortical afferents to the hippocampal formation in the rat. Neuroscience 4:463–476

Zimmer J, Tonder N, Sorensen T (1988) Hippocampus and fascia dentata transplants: Anatomical organization and connection. In: Chan-Palay V, Kohler C (eds) The Hippocampus: New vistas. Liss, New York (in press)

Can Fetal Cell Grafts be Expected to Ameliorate Symptoms of Human Neurodegenerative Disorders? Evidence from Animal Models

T. J. Collier, and *J. R. Sladek Jr.*

Summary

We are examining the therapeutic efficacy of replacement of neural tissue via transplantation in two animal models of human neurodegenerative disease:
1. the dopamine deficiency associated with MPTP treatment in monkeys, which reproduces many of the symptoms of Parkinson's disease, and
2. age-related deficits in brain norepiniphrine system function associated with memory deficits in old rats, a model of one of the changes seen in Alzheimer's disease.

Embryonic cell grafts of dopamine-producing neurons of the primate substantia nigra appear effective in reversing the motor deficits of experimental parkinsonism in 1-methyl-4phenyl-1,2,3,6-tetrahydropyridine (MPTP)-treated monkeys. Monkeys receiving dopamine-rich neural grafts have now been studied for over seven months after transplantation and the observations appear to support our earlier findings on the short-term therapeutic value of nerve cell grafts. In addition, this longer survival interval indicates that growth and integration of the grafted tissue with the host brain continues over time, with no signs of rejection of transplanted tissue. Our previous studies of aged rats indicate that a subpopulation of aged animals exhibit deficient memory for an inhibitory avoidance task and that one correlate of this age-related behavioral impairment is decreased norepinephrine content in the neurons of the locus coeruleus.

Supplementation of norepinephrine via implantation of fetal locus coeruleus neurons into the third ventricle reverses this behavioral deficit. We have recently shown that this same subpopulation of aged rats is deficient in learning and memory for a spatial localization task and that this behavior also is improved by implantation of norepinephrine-producing cells. Furthermore, evidence from a pilot study suggests that norepinephrine neurons grafted into young adult animals survive and grow over long intervals − 19 months − and appear to forestall the development of some age-related behavioral deficits. Thus, while significant questions remain concerning the mechanism of action of grafted neural tissue, evidence from animal models suggests that neural transplants can promote sustained behavioral improvement in the lesioned and aged central nervous system and provides hope that this technique may be of therapeutic value in some human neurodegenerative syndromes.

Introduction

Over the last twenty years an extensive series of experiments centering on a rodent model of the dopamine deficiency which produces the motor abnormalities of Parkinson's disease have suggested that behavioral recovery from this debilitating neurodegenerative syndrome may be achieved through replacement of dopamine neurons via intracerebral transplantation. Studies of nerve cell grafting in these dopamine-depleted rats indicated that developing dopamine neurons from the substantia nigra of fetal rat donors survive transplantation into the adult rat brain, continue to grow in the grafted site and send axons into the surrounding host brain, establish identifiable synaptic connections with host neurons, receive synaptic contacts from host neurons, continue to synthesize and release dopamine, exhibit spontaneous electrical activity reminiscent of normal nigral neurons, and perhaps most importantly, ameliorate the behavioral symptoms of dopamine depletion (Brundin et al. 1987; Dunnett et al. 1985; Freed et al. 1984). These landmark studies both mandated continued feasibility testing of dopamine neuron grafting in primate species as a prelude to clinical trials in human parkinsonism and raised questions related to the broader application of neural transplantation technology to a host of other neurodegenerative syndromes, in particular, the progressive degeneration associated with Alzheimer's disease. We are currently investigating the efficacy of dopamine neuron grafts in the MPTP toxicity model of Parkinson's disease in African green monkeys, and graft supplementation of one neurochemical deficit associated with cognitive decline in Alzheimer's disease that also occurs in aged rats – loss of norepinephrine in the pontine nucleus locus coeruleus.

Dopamine Neuron Grafts in Parkinsonian Monkeys

Systemic administration of the neurotoxin MPTP to African green monkeys produces degeneration of the nigrostriatal dopamine pathway yielding behavioral signs of human Parkinson's disease, including resting tremor, bradykinesia, and difficulty in initiating movements. The neuropathology and behavioral signs associated with MPTP treatment in this primate species provide a striking model of Parkinson's disease. In collaboration with D. E. Redmond, R. H. Roth, and colleagues at Yale University, we began to test the feasibility of dopamine neuron grafting in this primate model of parkinsonism. Our initial experiments had the modest goal of determining whether fetal primate dopamine neurons could survive intracerebral transplantation and whether these grafted neurons would have any discernible behavioral influence in three monkeys with mild to moderate motor symptoms induced by MPTP treatment. Indeed, this first study indicated that fetal monkey dopamine neurons of the substantia nigra, as well as other catecholamine-producing cells, survived transplantation, continued to express their neurotransmitter markers, and elaborated typical neuronal processes in the grafted site (Sladek et al. 1986, 1987b). In addition, dopamine-rich grafts placed into the appropriate target, the striatum, ameliorated behavioral signs of MPTP parkinsonism, while similar grafts placed inappropriately into the overlying cortex were without therapeutic effect (Redmond et al. 1986). Behavioral recovery was accompanied by increased levels of dopamine metabolites in the cerebrospinal fluid (Redmond et al. 1986; Sladek et al. 1987a). This initial study was carried out over the relatively short graft survival interval of 70 days. Thus, while feasibility of neural grafting in primates was clearly established, significant questions remained concerning the long-term efficacy of dopamine neuron implants, particularly in severely parkinsonian monkeys.

Recently, we have completed analysis of a set of three MPTP-treated monkeys that were behaviorally matched for severe parkinsonian symptoms and essentially immobile at the time of transplantation surgery (Sladek et al. 1988). These animals were followed for a period of 5−7.5 months after grafting. One animal received bilateral grafts of fetal substantia nigra dopamine-rich tissue into the striatum at three rostral-caudally spaced sites in each hemisphere. The second animal received the same number of grafts into the striatum but consisting of cerebellar tissue that contains no dopamine neurons. The third monkey received bilateral grafts of fetal nigral tissue misplaced into the overlying cortex. Only the animal receiving the appropriate dopamine neuron grafts into the striatum exhibited gradual behavioral improvement resulting in fully functional spontaneous behavior that remained stable 7.5 months after surgery. The other two animals remained severely debilitated and were sacrificed at shorter intervals (5 months) for humane reasons. Anatomical examination of their brains revealed surviving grafts in all three animals. The animal exhibiting behavioral recovery had clusters of grafted tyrosine hydroxylase positive neurons (marker of dopamine neurons) bilaterally in the striatum. Over this extended survival interval the grafted cells had elaborated long axonal processes that clearly extended across graft − host interfaces to innervate the dopamine-depleted host striatum. Biochemical analyses on tissue punches taken from striatum near to, and distant from, the dopamine neuron grafts indicated that grafts provided a local supplementation of dopamine that roughly corresponded to the anatomically identifiable sphere of graft-derived axonal influence. While the local recovery of dopamine levels was much below the levels normally found in the healthy striatum, the grafts did provide sufficient dopamine to correspond to values seen in MPTP-treated monkeys that do not show behavioral symptoms. Although the two nonresponsive control monkeys showed healthy grafts, it was clear that transplanted tissue that did not contain dopamine neurons (the cerebellar graft), or dopamine neurons placed outside the critical striatal target region (the graft into cortex), were not effective treatments.

We are continuing to study animals that have now hosted grafts for more than a year, and this second stage of studies has answered some significant questions. First, fetal neuron grafts do appear to be anatomically and functionally lasting for at least 7.5 months after implantation. Second, the longer survival interval has provided additional anatomical and biochemical evidence for continued integration of the graft with the host brain, both based on more extensive neurite outgrowth from grafted cells than was seen in the short-term grafts, and on the local changes in dopamine levels observed near the fetal neuron grafts. Third, the recovery of behavioral function appears to be dependent on replacement of the appropriate dopamine-producing cells into the appropriate striatal target. Any less specific influence of fetal tissue in general or misplaced dopamine grafts appears to be insufficient to provide a therapeutic influence in severely parkin-sonian monkeys. Finally, dopamine neuron grafts were effective in ameliorating behavioral symptoms in monkeys that were severely debilitated and essentially "end-stage" in their parkinsonism. Taken together, our evidence continues to sup-port the view that dopamine neuron replacement can provide behaviorally sig-nificant therapeutic benefit in a nonhuman primate model of Parkinson's disease, and encourages the continued development of this technique with an eye toward human application.

Norepinephrine Neuron Grafts in Aged, Memory-Deficient Rats

Application of neural grafting techniques to Alzheimer's disease presents an addi-tional set of problems. Even at the outset, the multifocal degeneration of Alz-heimer's disease complicates any approach such as neural grafting that seeks to replace neurotransmitters and circuitry. Whereas the lesion in Parkinson's disease that is primarily responsible for the motor deficits is loss of dopamine in the striatum, yielding the conceptually simple approach of grafting dopamine neurons back into the striatum, Alzheimer's disease is characterized by neuronal degener-ation in the nucleus basalis, locus coeruleus, hippocampus and cortex with accom-panying deficits in the neurotransmitters acetylcholine, norepinephrine, somato-statin, and others, all of which may contribute to the memory deficits and demen-tia that accompany the disease (Jellinger 1987). It is reasonable to question whether neural grafting can be of benefit in such a multisystem disorder. As a first approach to this problem, we and others have begun to examine grafting applied to individual neural deficits relevant to Alzheimer's disease.

The model system we have been utilizing is the aged (24-month-old) Fischer-344 rat. Behavioral testing of populations of aged rats reveals that a subpopulation of animals (approximately 40% of the population in our studies) exhibit marked deficits in memory for an inhibitory avoidance task (Collier et al. 1987). Memory-deficient aged rats in all probability experience age-related changes in multiple systems of the central nervous system. At least two such changes have been linked to deficient memory performance: changes in cholinergic (Bartus et al. 1982) and noradrenergic (Collier et al. 1987) system function. Both of these deficits are also seen in exaggerated form in Alzheimer's disease (Jellinger 1987). Our work has focused upon grafting norepinephrine neurons into memory-deficient aged rats. As reported previously (Collier et al. 1988), implantation of fetal norepinephrine neurons of the developing locus coeruleus into the third ventricle of aged rats ameliorates deficits in inhibitory avoidance memory performance. Furthermore, the therapeutic effect of the graft was blocked by pretreatment with the adrenergic receptor blocker propranolol, could be mimicked by intraventricular infusion of norepinephrine via mini-pump, and was not obtained with grafts of cerebellar tis-sue that contain no norepinephrine-producing cells. More recently, we have found that this subpopulation of aged rats is also deficient in learning and memory for a spatial localization task in a swim maze. This behavior appears to be less respon-sive to intraventricular grafts, but animals receiving norepinephrine neuron grafts into the cingulate cortex, a cortical region that lesion studies suggest is crucial for good performance of the spatial task (Sutherland et al. 1988), exhibit marked improvement in spatial memory (Fig. 1). Thus, as was the case with dopamine neuron grafts, it appears that norepinephrine neurons grafted in close proximity to brain regions involved in performance of a particular task can be of behavioral benefit. Furthermore, these grafted cells remain viable and exert their functional influence in an aged central nervous system.

In an initial study of the long-term viability of norepinephrine neuron grafts we implanted locus coeruleus tissue into the third ventricle of rats at 5 months of age, allowed these animals to grow old, and tested memory for the inhibitory avoidance task at 24 months of age. All six rats examined in this pilot study exhibited maximal retention of the task. While the number of animals studied is far too low to draw conclusions, it is impressive that none of these animals developed the memory deficit that would be expected in a portion of the rats this age. Anatomical analysis of these animals revealed large, organotypic grafts of locus coeruleus neurons in all six hosts. These intraventricular grafts had grown over the 19-month survival interval to reach a size that was two to three times that of grafts studied at a 2-month survival interval. Thus, norepinephrine neuron grafts survive and continue to grow over extended periods of time in aging rat hosts and may forestall the development of some age-related behavioral deficits that have been linked to aging changes in norepinephrine system function.

Can Neural Grafts Be Effective in Human Neurodegenerative Disease?

The question of applicability of neural grafting to human neurodegenerative disease is being answered in part by ongoing trials in human patients. A small number of individuals suffering from Parkinson's disease have been implanted with fetal

---

Fig. 1 A–C. An example of the morphology and behavioral effect of a fetal norepinephrine neuron graft placed into the posterior cingulate cortex of a 24-month-old rat. **A** Micrograph of a section through the posterior cingulate cortex processed for tyrosine hydroxylase immunocytochemistry. A cluster of darkly stained grafted fetal locus coeruleus neurons *(arrows)* is seen here in the midline cortex of the right hemisphere at six weeks posttransplantation. Grafted neurons and their processes were observed over a 2.5 mm rostral-caudal zone of cortex. *PCg,* posterior cingulate cortex; *Hpc,* dorsal hippocampus; *asterisk,* sagittal fissure; *arrowheads,* dorsal cortical surface. **B** Spatial localization performance in a Morris swim maze before grafting. The aged rat received 56 trials of training for localization of a hidden platform *(P)* maintained in a constant location in the pool. Twenty-four hours after the last training trial the platform was removed and the animal's swim pattern was recorded as a measure of retention of the trained location. As shown in the tracing presented, even after extensive training, this aged rat showed no focusing of its swim search near the previous location of the platform: an indication of impaired spatial memory. **C** Five weeks after norepinephrine neuron implantation, the same animal received 40 trials of training to a new spatial location. Again, 24 hours after the last training trial, a retention test was run. As shown in the swim-path tracing, the grafted aged animal exhibited improved spatial memory as evidenced by enhanced focusing of search in the vicinity of the platform location. Interestingly, the grafted animal also showed increased activity in the adjacent quadrant of the pool: the area that contained the platform during initial training five weeks previously. Thus, spatial memory appears to be improved for both the recently and previously learned locations

dopamine neurons by physicians in at least four groups around the world. All of these studies are considered to be in the early stages, and no conclusive information is currently available.

At the risk of stating the obvious, the predictive value of grafting experiments in animals for applications in human neurodegenerative diseases depends a great deal upon how closely the experimental animals model the human condition. Thus, highlighting some of the similarities and differences between the animal

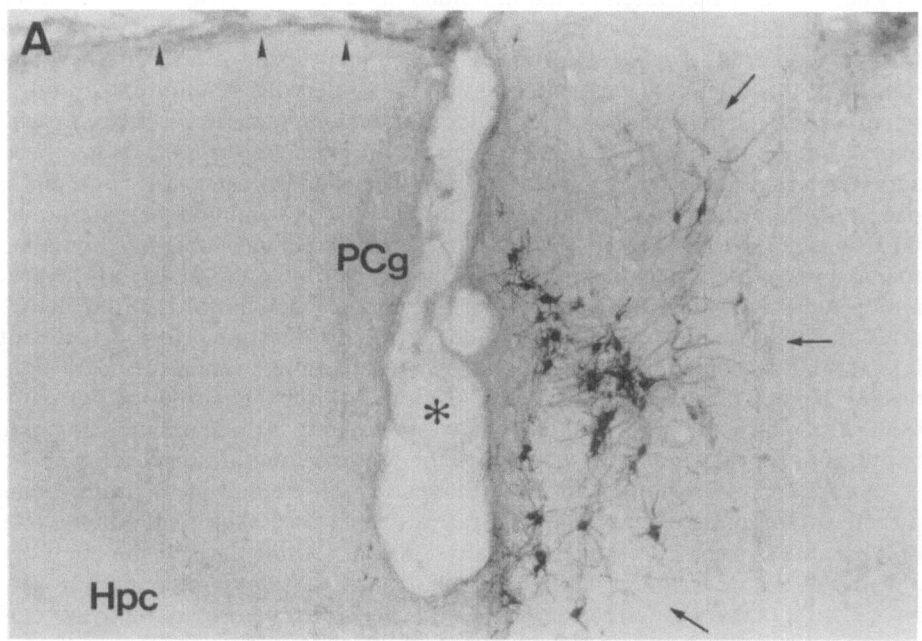

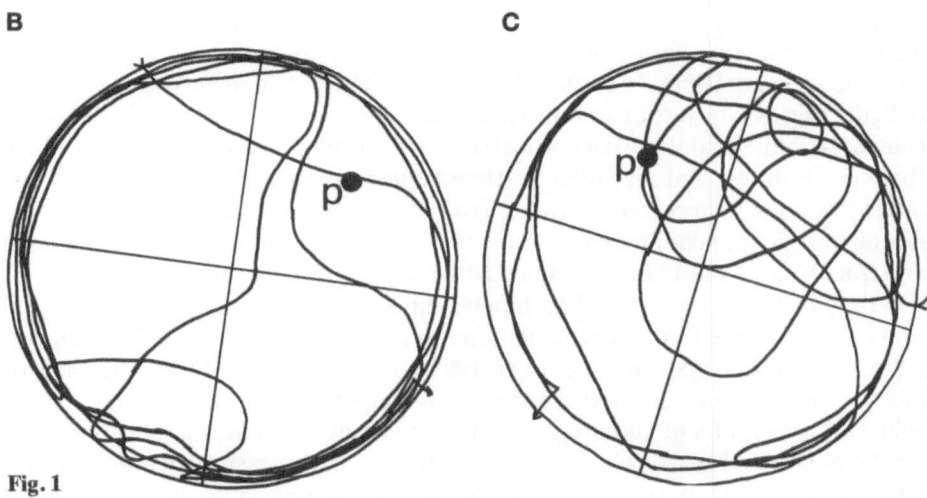

Fig. 1

models and the human conditions may be instructive. On a positive note, the results indicate that neural grafts can be effective in a central nervous system experiencing ongoing degeneration, as in the MPTP-treated monkeys, and in an aged nervous system, as in the rat experiments. Both progressive degeneration and old age are common characteristics of many human neurodegenerative syndromes that may be considered for treatment by neural grafting. The viability of grafted·neural tissue in aged and brain-damaged animal hosts is encouraging. A related issue is the timing of transplant intervention in animal models as compared to when it might be contemplated in human disease. For example, in our initial studies of parkinsonian monkeys, implantation of dopamine neurons was performed approximately 4 weeks after MPTP treatment. Is the efficacy of the grafts related to this relatively early intervention into a nervous system that is undergoing a complex combination of degeneration and compensation? If so, can we expect a similar efficacy of grafted neurons in a human brain that has been experiencing progressive degeneration over 10 or more years, as is common in patients with Parkinson's disease? Perhaps years of dopamine denervation produce changes in striatal target neurons that render them incapable of responding to dopamine replacement by grafted neurons (McNeill et al. 1988). Clearly, experiments in animals that have had their lesions for extended periods of time before grafting would be of great benefit in addressing this question. However, it is of interest that in our rat experiments, norepinephrine grafts retain their effectiveness even when placed into a behaviorally impaired animal near the end of its lifespan. These host animals are likely to have stable, long-term neural deficits prior to grafting, yet they remain responsive to the presence of grafted neurons.

As alluded to previously, human diseases that affect multiple neural systems provide additional problems in utilizing a graft-replacement approach. In the case of Alzheimer's disease, in which degeneration contributing to the dementia probably involves at least the basal forebrain cholinergic system, pontine noradrenergic cells, hippocampus, and cortex, the prospects for repairing all these systems appear grim. Indeed, as researchers have begun to test grafting in models of individual components of the disease, information that may reflect a basic limitation of the grafting approach appears to be accumulating. While grafts of either norepinephrine neurons (Collier et al. 1988) or acetylcholine neurons (Gage et al. 1984; Gage and Björklund 1986) have been effective in reversing behavioral deficits in aged rats, less functional success has been achieved with grafts of hippocampal (Kimble et al. 1986) or cortical (Gibbs et al. 1987) tissue. One way in which these grafts differ, that may relate to their behavioral efficacy, is the specificity of interaction that is necessary between grafted tissue and the host brain. Both the noradrenergic and cholinergic systems that exhibit functional decline in old age and frank degeneration in Alzheimer's disease provide widespread innervation of a variety of brain areas including hippocampus and cortex, and are believed to exert rather "global" influences on general excitability of their target neurons (Aston-Jones 1985; Aston-Jones et al. 1985). Thus, relatively diffuse presence of these neurotransmitters in the appropriate targets, as is likely to be readily achieved with grafts of cholinergic or noradrenergic neurons, may restore sufficient regulation of target cells to improve behavior. In contrast, both hippocampus and cortex are highly ordered brain structures that rely upon specific intrinsic organization and specific relationships with the targets they project upon for proper function. Reconstruction of such specific neural interactions, via transplantation of cortical or hippocampal tissue, does not as yet appear to be as readily achieved and indeed may be outside the practical limitations of the technique. However, it is interesting that grafting of either acetylcholine neurons or norepinephrine neurons into memory-impaired aged rats is sufficient to yield behavioral improvement in an animal that is almost certainly experiencing multiple neurochemical deficits. This evidence suggests that a therapeutic effect may be achieved by a partial supplementation of one or more compromised neural systems and need not require repair of all circuitry damaged by the disease.

Lastly, animal models cannot reproduce the etiology of human neurodegenerative syndromes. While some aspects of the pathological changes and symptoms can be mimicked, the animal models do not reproduce the progression of the disease process itself. Accordingly, it is unclear whether neural grafting can, indeed, yield sustained amelioration of behavioral deficits, or whether grafted cells would also undergo disease-related degeneration. This issue is unlikely to be answered until more information can be obtained from human clinical trials. It is encouraging, though, that our trials of noradrenergic grafts in young rats that were allowed to grow old survived and continued to exert a functional influence over the entire lifespan of the animal. While no known disease process was ongoing in these host animals, the grafted cells were exposed to progressive changes in the aging brain without apparent compromise of their functional capacity.

A recent modification of the neural grafting approach which promises to be of particular value in the treatment of neuronal degeneration syndromes is the use of grafted tissue, not as a replacement for lost or compromised neurons, but as a source of growth and survival factors that may stimulate repair and regeneration of the host brain. For example, a variety of evidence supports the view that basal forebrain cholinergic neurons are dependent upon the substance nerve growth factor (NGF) for their survival (Springer et al. 1987). Transection of their axons, interrupting the supply of NGF, leads to neuronal degeneration. In collaboration with Joe Springer of Hahnemann University we have shown that providing a tissue source of NGF, via intracerebral grafting of pieces of mouse submaxillary gland, can rescue and induce regeneration in axotomized cholinergic cells that would otherwise degenerate (Springer et al. 1988). Similarly, it has recently been demonstrated that fibroblasts that have been genetically altered to secrete NGF can be grafted to provide trophic support for axotomized cholinergic neurons (Rosenberg et al. 1988). This use of neural grafts to enhance the brain's own plasticity may have exciting applications in clinical neuroscience.

In conclusion, evidence based on neural grafting in animal models of human disease continues to provide hope that application of the transplant approach to some neurodegenerative syndromes may be of lasting therapeutic value. Ongoing research has progressed beyond questions of feasibility to test the practical limitations of the neural grafting approach. While work must continue on the mechanisms of graft–host interaction and the functional efficacy of neural grafts in animals with long-standing lesions, thus more accurately modeling the damage of human disease, it is clear that grafted neural tissue can be anatomically and functionally viable for long periods of time, even in an aging nervous system. The recent expansion of the neural grafting approach to include transplanted cells as sources of neuronal growth factors that may influence survival and repair of host neurons promises to widen the applicability of this technique.

Acknowledgments. The authors wish to acknowledge the important continuing contributions of Brian Daley and Barbara Blanchard to this work. This research was supported by NS24032 and an award from the Alzheimer's Disease and Related Disorders Association.

References

Aston-Jones G (1985) Behavioral functions of locus coeruleus derived from cellular attributes. Physiol Psychol 13:118−126

Aston-Jones G, Rogers J, Shaver RD, Dinan TG, Moss DE (1985) Age-impaired impulse flow from nucleus basalis to cortex. Nature 318:462−464

Bartus RT, Dean RL, Beer B, Lippa AS (1982) The cholinergic hypothesis of geriatric memory dysfunction. Science 217:408−417

Brundin P, Strecker RE, Lindvall O, Isacson O, Nilsson OG, Barbin G, Prochiantz A, et al. (1987) Intracerebral grafting of dopamine neurons. Experimental basis for clinical trials in patients with Parkinson's disease. In: Azmitia EC, Björklund A (eds) Cell and tissue transplantation into the adult brain. Ann NY Acad Sci 495:473−495

Collier TJ, Gash DM, Sladek JR (1987) Norepinephrine deficiency and behavioral senescence in aged rats. Transplanted locus coeruleus neurons as an experimental replacement therapy. In: Azmitia EC, Björklund A (eds) Cell and tissue transplantation into the adult brain. Ann NY Acad Sci 495:396−402

Collier TJ, Gash DM, Sladek JR (1988) Transplantation of norepinephrine neurons into aged rats improves performance of a learned task. Brain Res 448:77−87

Dunnett SB, Björklund A, Gage FH, Stenevi U (1985) Transplantation of mesencephalic dopamine neurons to the striatum of adult rats. In: Björklund A, Stenevi U (eds) Neural grafting in the mammalian CNS. Elsevier, Amsterdam, pp 451−469

Freed WJ, Hoffer BJ, Olson L, Wyatt RJ (1984) Transplantation of catecholamine-containing tissues to restore the functional capacity of the damaged nigrostriatal system. In: Sladek JR, Gash DM (eds) Neural transplants. Development and function. Plenum, New York, pp 373−406

Gage FH, Björklund A (1986) Cholinergic septal grafts into the hippocampal formation improve spatial learning and memory in aged rats by an atropine-sensitive mechanism. J Neurosci 6:2837−2847

Gage FH, Björklund A, Stenevi U, Dunnett SB, Kelly PAT (1984) Intrahippocampal septal grafts ameliorate learning impairments in aged rats. Science 225:533−536

Gibbs RB, Yu J, Cotman CW (1987) Entorhinal transplants and spatial memory abilities in rats. Behav Brain Res 26:29−35

Jellinger K (1987) Neuropathological substrates of Alzheimer's disease and Parkinson's disease. J Neural Transm 24:109−129

Kimble DP, Bremiller R, Stickrod G (1986) Fetal brain implants improve maze performance in hippocampal-lesioned rats. Brain Res 363:356−363

McNeill TH, Brown SA, Rafols JA, Shoulson I (1988) Atrophy of medium spiny I striatal dendrites in advanced Parkinson's disease. Brain Res 455:148−152

Redmond DE, Sladek JR, Roth RH, Collier TJ, Elsworth JD, Deutch AY, Haber S (1986) Fetal neuronal grafts in monkeys given methylphenyltetrahydropyridine. Lancet I:1125−1127

Rosenberg MB, Friedmann T, Robertson RC, Tuszynski M, Wolff JA, Breakfield XO, Gage FH (1988) Grafting genetically modified cells to the damaged brain: Restorative effects of NGF expression. Science 242:1575−1578

Sladek JR, Collier TJ, Haber SN, Roth RH, Redmond DE (1986) Survival and growth of fetal catecholamine neurons transplanted into primate brain. Brain Res Bull 17:809–818

Sladek JR, Collier TJ, Haber SN, Deutch AY, Elsworth JD, Roth RH, Redmond DE (1987a) Reversal of parkinsonism by fetal nerve cell transplants in primate brain. In: Azmitia EC, Björklund A (eds) Cell and tissue transplantation into the adult brain. Ann NY Acad Sci 495:641–657

Sladek JR, Redmond DE, Collier TJ, Haber SN, Elsworth JD, Deutch AY, Roth RH (1987b) Transplantation of fetal dopamine neurons in primate brain reverses MPTP induced parkinsonism. Prog Brain Res 71:309–323

Sladek JR, Redmond DE, Collier TJ, Blount J, Elsworth JD, Taylor J, Roth RH (1988) Fetal dopamine neural grafts: extended reversal of methylphenyltetrahydropyridine-induced parkinsonism in monkeys. In: Gash DM, Sladek JR (eds) Transplantation into the mammalian CNS. Prog Brain Res 78:497–506

Springer JE, Koh S, Tayrien MW, Loy R (1987) Basal forebrain magnocellular neurons stain for nerve growth factor receptor: Correlation with cholinergic cell bodies and effects of axotomy. J Neurosci Res 17:111–118

Springer JE, Collier TJ, Sladek JR, Loy R (1988) Transplantation of male mouse submaxillary gland increases survival of axotomized basal forebrain neurons. J Neurosci Res 19:291–296

Sutherland RJ, Whishaw IQ, Kolb B (1988) Contributions of cingulate cortex to two forms of spatial learning and memory. J Neurosci 8:1863–1872

Age-Related Changes in Rat Hippocampal Noradrenergic Transmission: Insights from *In Oculo* Transplants

B. Hoffer, P. Bickford-Wimer, M. Eriksdotter-Nilsson, A.-C. Granholm, M. Palmer, G. Gerhardt, L. Olson, Å. Seiger, and *G. Rose*

Summary

The hippocampus has been identified as a critical structure for learning and memory. Hippocampal dysfunction, especially in biogenic aminergic circuits, has been implicated with Alzheimer's disease. Age-related changes in the responsiveness of hippocampal pyramidal neurons to norepinephrine (NE) have been investigated here using electrophysiological techniques. Local application via microejection pressure in situ was employed to establish the dose which elicited a 50% change in spontaneous discharge rate of single pyramidal neurons; these data were used to construct dose–response curves for the population of neurons tested in rats 3–6, 18–20, and 27–30 months old. The percentage of cells responding in rats 18–20 and 27–30 months old decreased with NE, but no decrease in the population ED_{50} was observed. These data indicate that there is a progressive age-related decline in the postsynaptic response to NE in the rodent hippocampus in situ.

Intrinsic versus extrinsic determinants of these age-related alterations in hippocampal noradrenergic transmission were then investigated using intraocular allografts in rats. Three groups of animals were examined: young hippocampal transplants in young hosts, old transplants in old hosts, and young transplants in old hosts. Postsynaptic sensitivity to NE was measured by extracellular recordings of spontaneous activity and superfusion with known concentrations of catecholamines in the anterior chamber of the eye. Hill plots demonstrated that the dose–response relationships of NE-induced depressions were linear and parallel in three groups. Aged hippocampal grafts displayed a one order of magnitude subsensitivity to NE, which was highly significant. The EC_{50} for this group was 203.1 μM as compared to 29.2 μM in young grafts. Young intraocular grafts in old hosts responded similarly to young transplants in young hosts, with an EC_{50} of 32.4 μM for the depressant actions of NE. These changes were thus dependent on transplant age rather than host age, suggesting an involvement of intrinsic rather than extrinsic determinants in this model system. Taken together, these data support the concept that biogenic aminergic transmission deficits, which have been identified in Alzheimer's disease, may in part derive from an inherent subsensitivity which occurs during senescence.

Introduction

A decline in the ability to learn and remember new information is a nearly inevitable consequence of the aging process (Kubanis and Zornetzer 1981). In extreme cases, as can be seen in patients with Alzheimer's disease, the memory defect becomes severe enough to prevent independent functioning of the individual. To understand the basis for age-related memory dysfunction it is necessary to elucidate how the physiology of neurons in brain areas hypothesized to be important in memory changes during aging and what the factors are that precipitate these changes.

In humans, the hippocampal formation is known to play a critical role in the memory for new information (Scoville and Milner 1957; Weiskrantz an Warrington 1975; Zola-Morgan et al. 1982). In particular, the pyramidal neurons of regio superior have been shown to be necessary for this process (Squire 1986; Squire and Davis 1981). Recent evidence has also demonstrated the presence of morphological changes in the hippocampus even in the early stages of Alzheimer's disease (Ball et al. 1985; Geddes et al. 1985; Hyman et al. 1984). While the precise role of the hippocampus in learning and memory is still under investigation, considerable interest has focused on the cholinergic afferents to this structure. Cholinergic abnormalities are a hallmark of Alzheimer's dementia (Palacios 1982; Bartus et al. 1982; Perry et al. 1978; Drachman and Levitt 1974); in addition, the administration of cholinergic antagonists to both humans and experimental animals causes memory deficits that are comparable to those seen with aging (Bartus et al. 1982; Drachman and Levitt 1974).

However, age-related alterations in other memory-relevant neurotransmitter systems have also been reported. For example, NE has been shown to exert a modulatory influence on learning and memory (Gold and Zornetzer 1983; McGaugh et al. 1984; Squire and Davis 1981; Algeri et al. 1978; Arnsten and Goldman-Rakic 1985).

Deficits in central NE transmission may be responsible for declining cognitive performance during aging (Arnsten and Goldman-Rakic 1985). In addition, central NE is markedly reduced in Alzheimer's disease (Palmer et al. 1987; Gottfries 1982). Catecholaminergic pathways have also been shown to be altered during senescence using several in vitro biochemical indices. Various areas of the central nervous system (CNS) demonstrate decreases in catecholamine turnover (Osterburg et al. 1981) and tyrosine hydroxylase activity in aging (McGeer et al. 1971). Reductions in β-adrenergic receptor numbers have been reported (Misra et al. 1980; Pittman et al. 1980). Moreover, decreases in catecholamine-stimulated adenylate cyclase (Makman et al. 1980; Walker and Walker 1973) and cyclic nucleotide accumulation (Schmidt and Thornberry 1978) in old animals have been found. Finally, electrophysiological investigations have demonstrated an age-related reduction in the inhibitory potency of locally applied NE in the cerebellum (Bickford 1983; Marwaha et al. 1980), cingulate cortex (Jones and Olpe 1984), and neocortex (Jones and Olpe 1983). However, whether there are age-related alterations in NE actions in the hippocampus has not been clearly defined. Consequently, we examined postsynaptic sensitivity to NE exposure on pyramidal cell discharge rate. For in situ studies three different age groups of rats, 3−6 months,

18–20 months, and 27–30 months, were used with local application of NE from multibarrelled micropipettes.

A major issue in aging research is the differentiation of direct versus indirect sequelae of senescence. Towards these ends, use of *in oculo* grafts provides a unique opportunity to delineate intrinsic versus extrinsic determinants of age-related adrenergic deficits. Transplants of fetal central neurons permit the generation of "young-old chimeras." Thus, if the age-related biological properties of the graft maintain the donor-age timetable, intrinsic determinism would be postulated; in contrast, if the biological properties shifted to the host-age timetable, extrinsic influences would be suggested. This approach has been detailed elsewhere (Hoffer and Dunwiddie 1985).

Hippocampal grafts *in oculo* have been shown to survive and mature in an organotypic manner (Olson et al. 1977) and to receive a functional adrenergic innervation from the sympathetic ground plexus of the host iris (Freedman et al. 1979; Taylor et al. 1978). Such hippocampal grafts also manifest many physiological and pharmacological properties resembling hippocampus in situ (Freedman et al. 1979; Hoffer et al. 1977; Taylor et al. 1978).

The existence of a hippocampal noradrenergic pathway *in oculo* allowed us to directly examine the role of intrinsic versus extrinsic influences on age-related changes in hippocampal NE transmission. For this work three transplant experimental groups were also used, to allow independent manipulation of graft and host age and thus examine intrinsic versus extrinsic determinants of any possible age-related noradrenergic changes. In the first group (young/young) the grafts were examined 2–3 months after transplantation into the eyes of young adult (150 g) rats. The second group (old/old) was studied at a graft age of 22–23 months and a host age of 24–25 months. The third group (young/old) consisted of fetal hippocampal tissue transplanted to the anterior eye chamber of 20-month-old rats. Thus, at the time of recording the grafts were 2–3 months old and the hosts were 22–23 months old (Fig. 1). NE was administered by superfusion to the grafts.

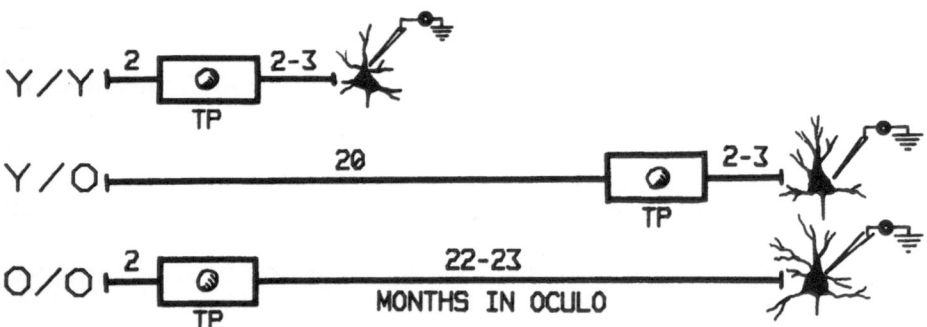

Fig. 1. Illustration of the experimental design in transplants. Three groups were studied. The *Y/Y* group (young grafts in young hosts) consisted of fetal hippocampal tissue transplanted *(TP)* to 2-month-old host rats which were thereafter allowed to mature 2–3 months before recording. The *Y/O* group (young grafts in old hosts) consisted of fetal hippocampal tissue transplanted to 20-month-old rats, which then remained in the eye 2–3 months before recording. Similarly, the *O/O* group (old grafts in old hosts) consisted of fetal hippocampal grafts, transplanted to 2-month-old rats, which then remained in the host eye for 22–23 months before recording

Results

Hippocampus In Situ

Pyramidal neurons in the hippocampal regio superior were identified by the duration of their unfiltered action potentials. Previous studies have demonstrated that a duration of 0.6–1.0 ms is characteristic of complex spike neurons (Rose et al. 1983). In addition, these neurons tend to discharge in bursts of three to five action potentials, termed complex spikes (Fox and Ranck 1981). The other major type of hippocampal neuron, the theta cell, has an action potential duration of less than or equal to 0.4 ms. Thus, these two types of hippocampal neurons are easily discriminated (Rose et al. 1983).

Local application of NE to complex spike neurons elicited a dose-dependent inhibition of firing rate (Fig. 2). By testing several, usually progressively increasing, doses of drug the dose (defined by the amount of pressure applied to the barrel in psi times the length of drug application in seconds, i.e., psi-sec) which elicited an approximately 50% inhibition in the firing rate of each neuron was determined. A curve fitting program was used to construct a cumulative dose–response curve for the populations of neurons examined in each age group (Fig. 3). This cumulative population dose–response analysis yielded an estimate of the percentage of neurons responding to locally applied monoamines and a population ED_{50} (i.e., the dose which elicited a response in 50% of the cells, excluding nonresponsive neurons). Both of these parameters were used to examine for possible age-related changes in response to locally applied NE.

Postsynaptic responsiveness to NE tended to decline with increasing age (Fig. 3). Analysis of the curves indicated that the 18 to 20-month and 27 to 30-month age groups were significantly different from the young animals with respect to the percent of cells responding, although no difference was observed for the population ED_{50} s (Table 1). The reduction in sensitivity to NE in the 18 to 20-month and 27 to 30-month age groups was confirmed by paired t-test ($n = 6$, $p < 0.05$ for both groups).

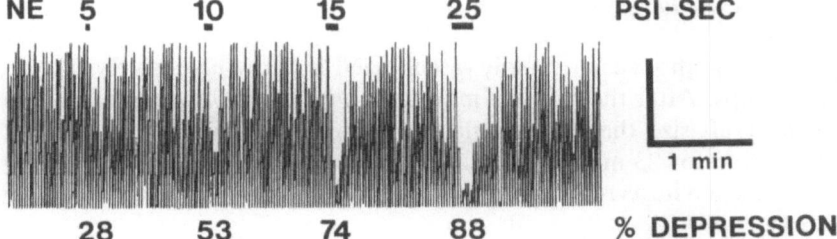

Fig. 2. Ratemeter record of a hippocampal CA1 complex spike neuron demonstrating a dose–response relationship to locally applied NE. A dose eliciting an approximately 50% inhibition was determined for each cell using this method. In this record time is indicated on the abscissa and the number of action potentials per second is indicated by the ordinate. *Bars* above the trace indicate the time when NE was being injected and the *numbers* above the *bar* specify the amount of NE in psi-sec. The percent depression of spontaneous firing rate elicited by NE application is indicated below the ratemeter record. The *vertical bar* represents 12 action potentials per second

NOREPINEPHRINE

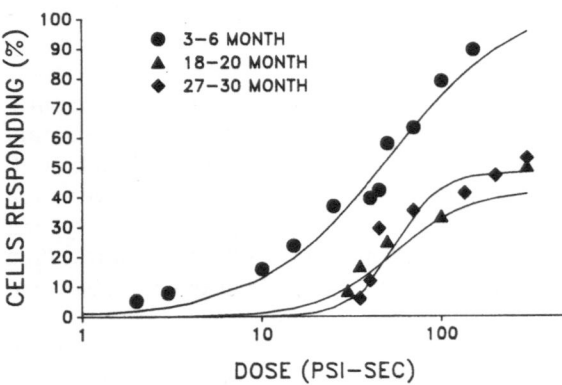

Fig. 3. Cumulative population dose response curves for the local application of NE demonstrating an age-related decline in responsivity. The ordinate represents the percentage of cells responding to local application of drug with a 50% decrease in spontaneous firing rate. The data are cumulative. The abscissa represents the dose of drug, in psi-sec, at or below which the cells responded to local application with a 50% inhibition. The number (n) of neurons recorded in each group were 3–6 months n = 36 cells, 20 rats; 18–20 months n = 13 cells, 8 rats; 27–30 months n = 15 cells, 5 rats

Table 1. Analysis of responses to locally applied norepinephrine in situ[a]

Age (months)	Percent responding ($\pm$ SEM)	Population ED_{50} ($\pm$ SEM)
3– 6	106.7 ± 5.8	51.6 ± 1.3
18–20	42.1 ± 2.9*	54.6 ± 1.6
27–30	45.3 ± 1.9*	55.2 ± 1.8

* Significantly different from 3–6 months, $p < .05$ and from 11–13 months, $p < .05$
[a] All values are calculated from the dose–response curves in Fig. 3

Hippocampus *In Oculo*

The hippocampal grafts became rapidly vascularized and grew extensively *in oculo* in all three groups. After the first few months *in oculo* the transplants ceased to grow and the graft size thereafter remained unchanged. The aged grafts that remained *in oculo* for 23 months did not decrease in size with age. The young transplants in the old hosts reached a smaller final size than young grafts in young rat hosts.

Extracellular recordings were performed from transplanted pyramidal cells, identified by their spontaneous complex spike discharge (see above). Discharge rates of individual neurons, tested with analysis of variance, did not significantly differ between the three groups. The average discharge rate in young grafts in young hosts was 4.4 ± 0.6 Hz (n = 15), and in old grafts in old hosts was 5.6 ± 0.5 Hz (n = 16). In young grafts in old hosts the discharge rate was 4.1 ± 0.4 Hz (n = 12).

Pyramidal neurons from young grafts in young hosts (young/young) responded to superfused NE with depressions in firing rates with an EC_{50} of 29.2 µM and a 95% confidence limit of 21.8−39.1 µM (Figs. 4 and 5). Pyramidal neurons in old hippocampal grafts (old/old) were significantly less sensitive than neurons in the young/young group, with a clear shift to the right in the dose−response curve for NE-induced depressions (Figs. 4 and 5). The EC_{50} was 203.1 µM with a 95% con-

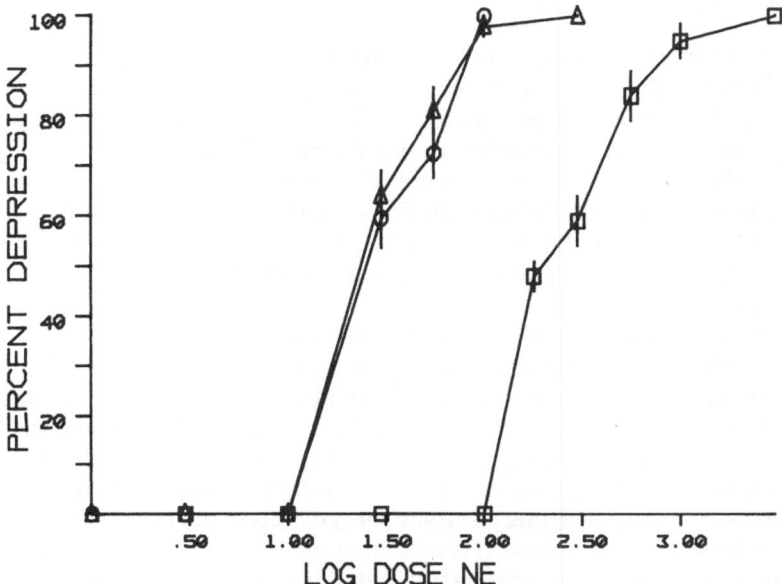

Fig. 4. Log dose−response curves for the depressant effects of superfused NE on the discharge rate of pyramidal neurons in Y/Y *(triangles)*, O/O *(squares)* and Y/O *(circles)* hippocampal transplants. Individual data points represent an *n* of 3−8 per point (mean ± SEM). The EC_{50} values and 95% confidence limits are calculated from linear (Hill) transformations

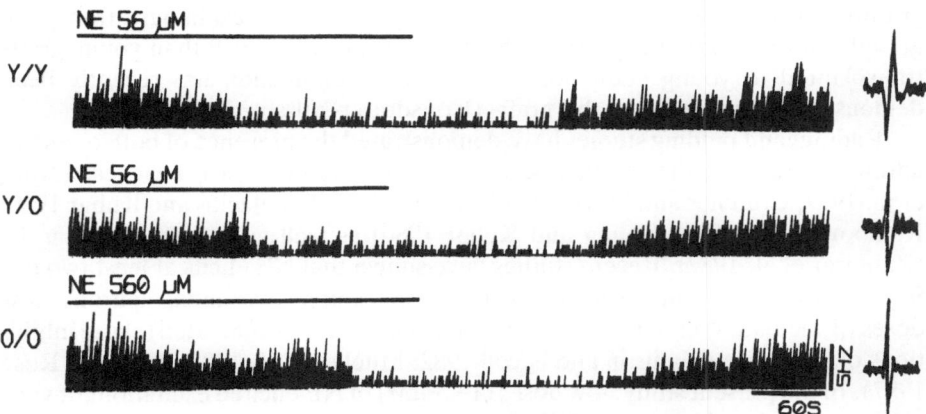

Fig. 5. Ratemeter records showing the depressant effects of superfused NE on hippocampal transplants. In the two young transplants (Y/Y and Y/O) 56 µM NE elicited a depression in neuronal activity. To obtain a similar degree of inhibition in an old graft (O/O), a concentration of 560 µM of superfused NE was needed

fidence limit of 162–254 μM. The NE-induced depressions in young grafts in old hosts (young/old) were almost identical to the inhibitions in the young/young group. The EC_{50} value was 32.4 μM with a 95% confidence limit of 25.5–41.3 μM (Figs. 4 and 5).

Discussion

A major initial finding of our experiments was the demonstration of an age-related decline in the postsynaptic sensitivity of in situ hippocampal pyramidal neurons to NE. The decrease in postsynaptic efficacy became apparent at 18–20 months. The postsynaptic response to NE declined further at 28–30 months of age.

The population dose–response curves obtained from the two oldest age groups showed significant alterations from those for the young (3 to 6-month) rats. The major component of this change was a downward shift resulting from a decrease in the percentage of responsive neurons in the hippocampus of older animals.

In this study we have also demonstrated parallel postsynaptic alterations in noradrenergic mechanisms in old hippocampal grafts *in oculo*. We found an age-related decline in noradrenergic receptivity of intraocular pyramidal neurons. The alterations of NE transmission seen here appear to be due to transplant age and not host age, suggesting an intrinsic determinism of NE mechanisms. These data are in line with a previous report from our laboratories where a subsensitivity to NE of Purkinje neurons was found in aged cerebellar grafts but not in young cerebellar grafts in old hosts (Granholm et al. 1987). This age-related decline in NE receptivity has also been demonstrated in situ in hippocampus here as well as in cerebellum, cingulate cortex, and neocortex (Bickford 1983; Bickford et al. 1985; Jones and Olpe 1983, 1984; Marwaha et al. 1980, 1981). Thus, a postsynaptic subsensitivity to NE may be a general phenomenon in the brain during aging.

Not all characteristics of a tranplant *in oculo* are independent of host age at grafting. Host age does seem to play a role regulating, at least in part, transplant growth. In this study, the grafts in the old hosts grew less well than young grafts transplanted to young adult hosts. A similar phenomenon has recently been demonstrated for cortex cerebri grafts (Eriksdotter-Nilsson et al. 1986).

Radioligand binding studies have demonstrated the presence of both α- and β-adrenergic receptors in the hippocampus in situ and in other brain areas (Atlas et al. 1977; Crutcher and Davis 1980; Jones et al. 1985; Palacios and Kuhar 1980; Rainbow et al. 1984; Young and Kuhar 1980) as well as in intraocular grafts (Zahniser et al. 1987). Recent studies have shown that NE elicits at least two different dose-dependent responses in the hippocampal pyramidal neurons; low doses (β-mediated) induce excitations while high doses (α-mediated) cause inhibitions of discharge (Madison and Nicoll 1982; Mueller et al. 1982; Pang and Rose 1987). In the present study, low doses (1–5 μM) of NE elicited excitations of variable magnitude which were difficult to quantitate but did not differ markedly between the three groups. However, higher doses of NE elicited depressions of pyramidal neuron firing rate. The aged hippocampal transplants, like hippocampus in situ, were clearly subsensitive to NE in their inhibitory response, with dose—response curves shifted to the right by one order of magnitude. These data extend biochemical studies which indicate that both α- and β-adrenergic receptor affinity or density are decreased in the brain during aging (Greenberg and Weiss 1978; Leslie et al. 1985; Maggi et al. 1979; Misra et al. 1980; Pittman et al. 1980). It must be emphasized that the functional NE subsensitivity reported here may not be causally related to changes in receptor number or affinity. Transduction mechanisms, such as G proteins, adenylate cyclase activity, phosphatidylinositol turnover, intracellular Ca^{2+} levels, and various protein kinases could certainly be major loci of age-related changes in noradrenergic transmission.

The relationship between the degeneration or slowing of locus coeruleus neurons in the senescent rats and the postsynaptic NE subsenstivity reported here and by other investigators is not clear. It might be expected that with the loss of NE input a "disuse" supersensitivity rather than subsensitivity would occur. Even though both NE content and turnover have been shown to be decreased in brains of aged rodents and primates (Arnsten and Goldman-Rakic 1985; Estes and Simpkins 1980; Finch 1973; Goldman-Rakic and Brown 1981; Gottfries 1982; McGeer et al. 1971; Winblad et al. 1985) plasma NE levels of aged Fisher-344 rats are increased (Chiueh et al. 1980). This may be indicative of an increased activity of the adrenals and the sympathetic ganglia during senescence, perhaps related to receptor subsensitivity.

Recent studies have suggested a significant relationship between loss of central noradrenergic function and senescent memory decline (Gage and Björklund 1986; Leslie et al. 1985). Leslie et al. (1985) showed a clear correlation between cell loss in locus coeruleus and retention latency on an inhibitory avoidance task in aged mice. Clinical studies have also shown a relationship between noradrenergic impairment and senescent memory decline (Bondareff et al. 1982; Iversen et al. 1983). Thus, although age-related neuronal loss has been demonstrated in several transmitter systems in the aged brain (Brizzee and Ordy 1979; Brody 1976; Ellis 1920; Landfield et al. 1977), degeneration of the locus coeruleus noradrenergic neurons has been specifically associated with senescent memory loss. A decreased activity or loss of neurons displaying higher firing rates has also been reported in the locus coeruleus of aged rats (Olpe and Steinmann 1982; Vijayashankar and Brody 1979; Wree et al. 1980). Prolonged electrical stimulation of the locus coeruleus or disinhibition with α-adrenergic receptor antagonists has also been reported to prevent age-related memory deficits without affecting behavioral response in young animals (Zornetzer 1985).

In conclusion, the present paper provides further evidence for a decline of central noradrenergic functions during senescence. A significant age-related decrease in the ability of NE to reduce the spontaneous activity of hippocampal pyramidal neurons was observed. Studies using *in oculo* grafts suggest this subsensitivity is related to factors intrinsic to the hippocampus.

References

Algeri S, Bonati M, Brunello N, Ponzio F (1978) Biochemical changes in central catecholaminergic neurons of the senescent rat. In: Deniker P, Radouco-Tomas C, Villeneuve A, Baronet-LaCroix C, Garcin F (eds) Neuropsychopharmacology. Proceedings of the tenth congress of the Colloquium International Neuro-Psychopharmacologicum, vol II, workshop 6: models in geriatric neuropsychopharmacology. Pergamon, Oxford, pp 1647–1654

Arnsten AFT, Goldman-Rakic PS (1985) Alpha$_2$-adrenergic mechanisms in prefrontal cortex associated with cognitive decline in aged nonhuman primates. Science 230:1273–1276

Atlas D, Teichberg VI, Changeus JP (1977) Direct evidence for beta-adrenoceptors on the Purkinje cells of mouse cerebellum. Brain Res 128:532–536

Ball MJ, Hachinski V, Fox A, Kirshen AJ, Fiman M, Blume W, Krai VA, Fox H (1985) A new definition of Alzheimer's disease: a hippocampal dementia. Lancet I:14–16

Bartus RT, Dean RL, Beer B, Lippa AS (1982) The cholinergic hypothesis of geriatric memory dysfunction. Science 217:408–417

Bickford PC (1983) Age-related alterations in noradrenergic neurotransmission in Sprague-Dawley and Fischer 344 rat strains. Age 6:100–105

Bickford PC, Hoffer BJ, Freedman R (1985) Interaction of norepinephrine with Purkinje responses to cerebellar afferent inputs in aged rats. Neurobiol Aging 6:89–94

Bondareff W, Mountjoy CQ, Roth M (1982) Loss of neurons of origin of the adrenergic projection to cerebral cortex (nucleus locus coeruleus) in senile dementia. Neurology (NY) 32:164–168

Brizzee KR, Ordy JM (1979) Age pigments, cell loss and hippocampal function. Mech Ageing Dev 9:143–162

Brody H (1976) An examination of cerebral cortex and brain stem aging. In: Terry RD, Gershon S (eds) Neurobiology of aging. Raven, New York, pp 177–181

Chiueh CC, Nespor SM, Rapoport SL (1980) Cardiovascular, sympathetic and adrenal cortical responsiveness of aged Fischer 344 rats to stress. Neurobiol Aging 1:157–163

Crutcher KA, Davis JN (1980) Hippocampal alpha- and beta-adrenergic receptors: comparison of (3H)dihydroalprenol and (3H) WB 4101 binding with noradrenergic innervation in the rat. Brain Res 182:107–117

Drachman DA, Levitt JB (1974) Human memory and the cholinergic system: a relationship to aging. Arch Neurol 30:113–121

Ellis RS (1920) Norms for structural changes in the human cerebellum from birth to old age. J Comp Neurol 32:1–34

Eriksdotter-Nilsson M, Björklund H, Dahl D, Olson L (1986) Growth and development of intraocular fetal cortex cerebri grafts in rats of different ages. Dev Brain Res 28:75–84

Estes KS, Simpkins JW (1980) Age-related alterations in catecholamine concentrations in discrete preoptic area and hypothalamic regions in the male rat. Brain Res 194:556–560

Finch CE (1973) Catecholamine metabolism in the brain of aging male mice. Brain Res 52:261–276

Fox SE, Ranck JB (1981) Electrophysiological characteristics of hippocampal complex-spike cells and theta cells. Exp Brain Res 41:399–410

Freedman R, Tailor D, Seiger Å, Olson L, Hoffer B (1979) Seizures and related epileptiform activity in hippocampus transplanted to the anterior chamber of the eye: modulation by cholinergic and adrenergic input. Ann Neurol 6:281–295

Gage FH, Björklund A (1986) Neural grafting in the aged rat brain. Annu Rev Physiol 48:447–459

Geddes JW, Monaghan DT, Cotman CW, Lott IT, Kim RC, Chui HC (1985) Plasticity of hippocampal circuitry in Alzheimer's disease. Science 230:1179–1181

Gold PE, Zornetzer SF (1983) The mnemon and its juices: neuromodulation of memory processes. Behav Neurol Biol 38:151–189

Goldman-Rakic PS, Brown RM (1981) Regional changes of monoamines in cerebral cortex and subcortical structures of aging rhesus monkeys. Neuroscience 6:177–187

Gottfries CG (1982) The metabolism of some neurotransmitters in ageing and dementia disorders. Gerontology 28:11–19

Govoni S, Memo M, Saiani L, Spano PF, Trabucchi M (1980) Impairment of brain neurotrans-
  mitter receptors in aged rats. Mech Ageing Dev 12:39–46
Granholm AC, Gerhardt GA, Eriksdotter-Nilsson M, Bickford-Wimer PC, Palmer MR, Seiger
  Å, Olson L, Hoffer BJ (1987) Age-related changes in cerebellar noradrenergic pre- and post-
  synaptic mechanisms: intrinsic vs extrinsic determinants evaluated with brain grafts in oculo.
  Brain Res 423:71–78
Greenberg LH, Weiss B (1978) Beta-adrenergic receptors in aged rat brain: reduced number
  and capacity of pineal gland to develop supersensitivity. Science 201:61–63
Hoffer BJ, Dunwiddie TV (1985) Brain grafts: potential therapy in neurodegenerative diseases
  and in understanding normal aging in the CNS. Neurobiol Aging 6:162–163
Hoffer BJ, Seiger Å, Taylor D, Olson L, Freedman R (1977) Seizures and related epileptiform
  activity in hippocampus transplanted to the anterior chamber of the eye. I. Characterization
  of seizures, interictal spikes and synchronous activity. Exp Neurol 54:233–250
Hyman BT, van Hoeven GW, Damasis AR, Branes CS (1984) Alzheimer's disease: cell specific
  pathology isolates the hippocampal formation. Science 235:1168–1170
Iversen LL, Rossor MN, Reynolds GP, Hills R, Roth M, Mountjoy CQ, Foote SL, Morrison
  JH, Blom FE (1983) Loss of pigmented dopamine-β-hydroxylase positive cells from locus
  coeruleus in senile dementia of Alzheimer's type. Neurosci Lett 39:95–100
Jones LS, Gauger LL, Davis JN (1985) Anatomy of brain alpha₁-adrenergic receptors: in vitro
  autoradiography with (125I)-HEAT. J Comp Neurol 231:190–208
Jones RSG, Olpe HR (1983) Altered sensitivity of forebrain neurons to iontophoretically
  applied noradrenaline in aging rats. Neurobiol Aging 4:97–99
Jones RSG, Olpe HR (1984) Multiple changes in the sensitivity of cingulate cortical neurons to
  putative neurotransmitters in aging rats: substance P, acetylcholine and noradrenaline.
  Neurosci Lett 50:31–36
Kubanis P, Zornetzer SF (1981) Age related behavioral and neurobiological changes: a review
  with emphasis on memory. Behav Neural Biol 31:115–172
Landfield PW, Rose G, Sandles L, Wohlstadter TC, Lynch G (1977) Patterns of astroglial
  hypertrophy and neuronal degeneration in the hippocampus of aged, memory-deficient rats.
  J Gerontol 32:3–12
Leslie FM, Loughlin SE, Sternberg DB, McGauch JL, Young L, Zornetzer SF (1985) Nor-
  adrenergic changes and memory loss in aged mice. Brain Res 359:292–299
Madison DV, Nicoll RA (1982) Noradrenaline blocks accommodation of pyramidal cell dis-
  charge in the hippocampus. Nature 299:636–638
Maggi A, Schmidt MJ, Ghetti B, Enna SJ (1979) Effect of aging on neurotransmitter receptor
  binding in rat and human brain. Life Sci 24:367–374
Makman MH, Ahn HS, Thal TJ, Sharpless NS, Dvorkin B, Horowitz SG, Rosenfeld M (1980)
  Evidence for selective loss of brain dopamine-histamine-stimulated adenylate cyclase
  activities in rabbits with aging. Brain Res 192:177–183
Marwaha J, Hoffer B, Pittman R, Freedman R (1980) Age-related electrophysiological changes
  in rat cerebellum. Brain Res 20:85–97
Marwaha J, Hoffer BJ, Freedman R (1981) Changes in noradrenergic neurotransmission in rat
  cerebellum during aging. Neurobiol Aging 2:95–98
McGaugh JL, Liang KC, Bennett C, Sternberg DB (1984) Adrenergic influences on memory
  storage: interaction of peripheral and central systems. In: Lynch G, McGaugh JL, Wein-
  berger NN (eds) Neurobiology of learning and memory. Guilford, New York, pp 313–332
McGeer EG, Fibiger HC, McGeer PL, Wickson V (1971) Aging and brain enzymes. Exp
  Gerontol 6:391–396
Misra CH, Shelat HS, Smith RC (1980) Effect of age on adrenergic binding in rat brain. Life Sci
  27:521–526
Mueller AL, Palmer MR, Hoffer BJ, Dunwiddie TW (1982) Hippocampal noradrenergic
  responses in vivo and in vitro: characterization of alpha and beta components. Arch Pharmacol 318:259–166
Olpe HR, Steinmann MW (1982) Age-related decline in the activity of noradrenergic neurons
  of the rat locus coeruleus. Brain Res 251:174–176

Olson L, Freedman R, Seiger Å, Hoffer B (1977) Electrophysiology and cytology of hippocampal formation transplants in the anterior chamber of the eye. I. Intrinsic organization. Brain Res 119:87–106

Osterburg HH, Donahue HG, Severson JA, Finch CE (1981) Catecholamine levels and turnover during aging in brain regions of male C57BL/6J mice. Brain Res 224:337–352

Palacios JM (1982) Autoradiographic localization of muscarinic cholinergic receptors in the hippocampus of patients with senile dementia. Brain Res 243:173–175

Palacios JM, Kuhar MJ (1980) Beta-adrenergic receptor localization by light microscopic autoradiography. Science 208:1378–1380

Palmer AM, Wilcock GK, Esire MM, Francis PT, Bowen DM (1987) Monoaminergic innervation of the frontal and temporal lobes in Alzheimer's disease. Brain Res 401:231–238

Pang K, Rose GM (1987) Differential effects of norepinephrine on hippocampal complex-spike theta neurons. Brain Res 425:146–158

Perry EK, Tomlinson BE, Blessed G, Bergmann K, Gibson PH, Perry RH (1978) Correlation of cholinergic abnormalities with senile plaques and mental test scores in senile dementia. Br Med J 2:1457–1459

Pittman RN, Minneman KP, Molinoff PB (1980) Alteration in beta$_1$- and beta$_2$-adrenergic receptor density in the cerebellum of aging rats. J Neurochem 35:273–275

Rainbow TC, Parsons B, Wolfe BB (1984) Quantitative autoradiography of beta$_1$- and beta$_2$-adrenergic receptors in the rat brain. Proc Natl Acad Sci USA 81:1585–1589

Rose G, Diamond D, Lynch GS (1983) Dentate granule cells in the rat hippocampal formation have the behavioral characteristics of theta neurons. Brain Res 266:29–37

Schmidt MJ, Thornberry JF (1978) Cyclic AMP and cyclic GMP accumulation in vitro in brain regions of young, old and aged rats. Brain Res 139:169–177

Scoville WB, Milner J (1957) Loss of recent memory after bilateral hippocampal lesions. J Neurol Neurosurg Psychiatry 20:11–21

Squire LR (1986) Mechanisms of memory. Science 232:1612–1619

Squire LR, Davis HP (1981) The pharmacology of memory: a neurobiological perspective. Annu Rev Pharmacol Toxicol 21:323–356

Taylor D, Seiger Å, Freedman R, Olson L, Hoffer B (1978) Electrophysiological analysis of functional reinnervation of transplants in the anterior chamber of the eye by the autonomic ground plexus of the iris. Proc Natl Acad Sci USA 75:1009–1012

Vijayashankar N, Brody H (1979) A quantitative study of the pigmented neurons in the nuclei locus coeruleus and subcoeruleus in man as related to aging. J Neuropathol Exp Neurol 38:490–494

Walker JB, Walker JP (1973) Properties of adenylate cyclase from senescent rat brain. Brain Res 54:391–396

Weiskrantz L, Warrington EK (1975) The problem of the amnesic syndrome in man and animals. In: Isaacson RL, Pribram KH (eds) The hippocampus, vol 2. Plenum, New York, pp 411–428

Winblad B, Hardy J, Backman L, Nilsson L-G (1985) Memory function and brain biochemistry in normal aging and in senile dementia. Ann NY Acad Sci 444:255–268

Wree A, Braak H, Schleicher A, Zilles K (1980) Biomathematical analyses of the neuronal loss in the aging human brain of both sexes, demonstrated in pigment preparation of the pars cerebellaris loci Coerulei. Anat Embryol (Berl) 160:105–119

Young WS, Kuhar MJ (1980) Noradrenergic alpha$_1$- and alpha$_2$-receptors: light microscopic autoradiographic localization. Proc Natl Acad Sci USA 77:1696–1700

Zahniser NR, Curella P, Burnett DM, Miller JA, Eriksdotter-Nilsson M, Granholm A-C (1987) Quantitative autoradiographic analysis of alpha$_1$- and beta-adrenergic receptors in intraocular rat cerebellar grafts. Soc Neurosci Abstr 13:1338

Zola-Morgan S, Squire LR, Mishkin M (1982) The neuroanatomy of amnesia: amygdala versus temporal stem. Science 218:1337–1339

Zornetzer SF (1985) Catecholamine system involvement in age-related memory dysfunction. Ann NY Acad Sci 444:242–254

Neural Grafts and Neurotransmitter Interactions in Cognitive Deficits

M. Segal, G. Richter-Levin, V. Greenberger, and R. Shpiegelman

Summary

Serotonin is a major modulatory neurotransmitter in the brain. Activation of serotonin receptors causes suppression of spontaneous activity in the hippocampus and an enhancement of evoked responses to afferent stimulation. Depletion of serotonin reduces the ability of the hippocampus to express long-term potentiation of the responses to afferent stimulation following a tetanic stimulation. Serotonin depletion by itself does not affect spatial memory ability of rats but when combined with a partial suppression of cholinergic neurotransmission there is a marked impairment of the ability to negotiate a spatial memory task.

A serotonin-containing transplant can incorporate into a serotonin-depleted host hippocampus; the grafted serotonin neurons will form connections with the host hippocampus and activation of the graft will produce effects in the host hippocampus akin to those produced in normal, but not in serotonin-depleted, brains. A serotonin-containing raphe graft can ameliorate cognitive deficits caused by a combined interruption of cholinergic and serotonergic neurotransmission. Thus, serotonin can interact with the cholinergic system to regulate cognitive functions associated with the hippocampus. This interaction is likely to take place in the hippocampus.

Introduction

The original description of severe cholinergic deficits in senile dementia of the Alzheimer's type (Alzheimer's disease, AD) (Davies and Maloney 1976; Perry et al. 1978) triggered a large number of studies in which cholinergic replacement therapy was attempted to ameliorate the cognitive deficits associated with AD (Barbeau et al. 1979; Corkin et al. 1982). The analogy with another neurodegenerative disease, parkinsonism, encouraged such an approach; as in Parkinson's disease (PD), AD was associated with a selective degeneration of specific basal forebrain acetylcholine-containing nuclei. In PD the main current effective treatment is to enhance formation of dopamine by supplying its precursor L-dopa. Presumably L-dopa enters the few remaining dopaminergic terminals and increases the amount of released dopamine. Using the same strategy, attempts were made to feed AD patients with large amounts of choline, a precursor for the depleted acetylcholine (ACh), or physostigmine which prevents the breakdown of ACh, or other drugs which interact with the cholinergic system (Corkin et al. 1982). This strategy is supported by a large body of evidence for the involvement of ACh in mnemonic functions of the brain (Deutsch 1983).

Unfortunately, none of these treatments has proved effective to date. In searching for possible causes for this apparent failure it becomes evident that in the AD brain the cholinergic system is not the only one to degenerate and that other neurotransmitter systems, including the monoamines, are also markedly impaired compared to those of a normal brain (Rossor and Iversen 1986).

This brings up the possibility that the severe cognitive deficits seen in AD patients may result from a combined loss of cholinergic and monoamine-containing neurons. If so, a more efficient therapy may require enhancement of both cholinergic and monoamine neurotransmission. To examine these possibilities one needs to revert to experimental models where both the combined damage and combined therapy can be studied in controlled conditions.

One of the new approaches to treatment of neurodegenerative disorders involves grafting of embryonic neural tissue enriched with the neuron type that is missing in the impaired adult brain. This approach has been used successfully in a number of animal models, and the general impression is that grafting can restore impaired neurological functions. Numerous examples of this generalization include animal models of Parkinsonism: in the unilateral nigrostriatal-lesioned rat and the methyl-4-phenyl-1,2,3,6-tetrahydropyridine (MPTP)-treated monkey grafting of embryonic midbrain neurons restores nearly normal motor function (Freed 1983; Stenevi et al. 1976; Dunnett et al. 1982; Strömberg et al. 1985). Recent attempts to use neural grafts to treat PD are encouraging but attest to the need for an extensive understanding of the graft−host interaction before such an attempt is successful. Several recent studies suggest that cognitive deficits in animal models of AD following chemical lesions of basal forebrain cholinergic neurons can be restored following injection of basal forebrain neurons into the neocortex or the hippocampus (Dunnett et al. 1982). The rules governing involvement of basal forebrain cholinergic neurons in cognitive functions, as well as the mechanisms underlying the restoration of functions following transplantation, are not entirely clear and further basic studies are needed before treatment of AD with grafted cholinergic neurons can be attempted.

One of the neurotransmitters impaired in AD patients is the midbrain serotonergic system (Rossor and Iversen 1986). Depletion of serotonin in rats does not produce any clear cognitive deficit (unpublished observations). However, it is quite possible that a combined neurotransmitter loss will have an interactive effect on cognitive functions. Assuming that cholinergic and serotonergic neurotransmission interact in controlling cognitive functions (Vanderwolf 1987) and that severe cognitive deficits can result from a combined loss of cholinergic and serotonergic neurotransmission, we searched for possible combined roles of serotonin and acetylcholine in plastic functions in the brain. The experiments described in this chapter address the following questions: How does serotonin, released from its terminals in the hippocampus, affect the spontaneous, evoked, and plastic properties of a specific sector of the hippocampus, the dentate gyrus? Next we explore the possibility that grafted raphe neurons can incorporate into the hippocampal circuit and restore functions lost by a previous chemical lesion of the native serotonergic system. Finally, we examine the possible interaction between serotonergic and cholinergic neurotransmission and the possible amelioration of cognitive deficits caused by a combined lesion, using raphe or raphe/septal graft.

Physiological Studies

Serotonin-containing midbrain raphe neurons project diffusely throughout the neural axis. The hippocampus is one of the main terminal areas of raphe fibers and has one of the highest concentrations of serotonin receptors. Topical application of serotonin onto hippocampal neurons in vivo produced inhibition of spontaneous action potential discharges (Segal 1975) and a hyperpolarization associated with an increase in potassium conductance when studied intracellularly in an in vitro slice preparation (Segal 1980). In addition, serotonin blocks a slow afterhyperpolarization and can produce a slow, delayed depolarization (Andrade and Nicoll 1987; Colino and Halliwell 1987). These two effects of serotonin can cause an increase in excitability of hippocampal neurons. In the intact brain it has been shown (Assaf and Miler 1978; Winson 1980) that electrical stimulation of the raphe nuclei can enhance reactivity of the dentate gyrus to stimulation of a major excitatory afferent arriving from the entorhinal cortex via the perforant path. Experiments using electrical stimulation of the raphe are confounded by the heterogeneity of this nucleus and the fact that it contains many nonserotonergic cells and fibers of passage. We have used d-fenfluramine (FFA), a drug known to release 5-HT from its terminals, to study the role of serotonin in regulating dentate gyrus excitability. FFA caused a marked (50%–100%) increase in population spike response to perforant path stimulation without causing any systematic change in population excitatory postsynaptic potential (EPSP) response to the same stimulation (Richter-Levin and Segal 1988). This indicates that FFA affects primarily dentate cell excitability, i.e., the input/output ratio in these cells. Further experiments indicated that serotonin released by FFA modulates a feed-forward inhibitory control of dentate activity arising in the contralateral hippocampus.

The role of serotonin in regulating plastic properties of the dentate gyrus was examined using tetanic stimulations of the perforant path. Such a stimulation causes a long-term potentiation of the reactivity of the dentate gyrus to this stimulation. This well-known long-term potentiation (LTP) has been studied extensively in the dentate gyrus (Bliss and Dolphin 1982). We found that serotonin-depleted rats exhibit less LTP than do normal control rats. This is perhaps due to a reduced ability to regulate feed-forward inhibition in these serotonin-deficient hippocampi.

A major disadvantage of the study of drug effects on physiology and behavior is that the drug may cause side effects not directly relevant to the action of interest. One rationale for using brain grafts is that they may help in directing drug action to specific regions of interest in the brain. For example, in a serotonin-depleted brain we can expect FFA to release serotonin only from the region innervated by the graft.

We examined the possible incorporation of raphe grafts in the hippocampus and the possible restoration of putative serotonergic functions in these brains. Initially we found that serotonergic neurons grafted into the hippocampus develop normal physiological properties akin to those of normal in situ serotonergic neurons (Segal 1987; Segal and Azmitia 1986). These studies were done in the slice preparation which contained both grafted and host neurons. We also found that the grafted neurons innervate the host hippocampus and, when stimulated, produce a typical hyperpolarizing response in host neurons (Fig. 1). Thus, it appears that the graft is physiologically incorporated in the host hippocampal circuitry. We pursued our studies into the intact brain and found that FFA can produce, in a hippocampus grafted with serotonin neurons, the typical enhancement of reactivity to perforant path stimulation (Segal et al. 1988; Fig. 2). Furthermore, the graft enhanced the ability of the hippocampus to express LTP to perforant path stimulation.

One of the main effects of FFA is to produce a marked suppression of spontaneous EEG spikes and high voltage wave activity. This effect was seen in a normal brain and in brains grafted with midbrain raphe neurons but not in serotonin-depleted brains. The increased reactivity to afferent stimulation along with a decrease in spontaneous activity underlie the unique nature of neuromodulation exerted by serotonin in the hippocampus.

Behavioral Studies

Normal adult rats can learn to find a submerged platform in a pool (i.e., water maze) of water using spatial, extra-pool cues (Morris 1984). Once reaching the platform they can escape the water maze. Hippocampectomized rats or rats in which the hippocampus was disconnected from the rostral forebrain by a fornix transection are unable to solve the water maze problem using spatial cues. We found that suppression of cholinergic muscarinic neurotransmission with a high dose (50 mg/kg) of the muscarinic antagonist atropine can also block performance of rats in the water maze. Such a high dose causes the rats to shift from a spatially cued search for the platform to a random strategy of platform search. A lower dose (20 mg/kg i.p.) did not impair the ability of the rats to successfully navigate the water maze and they were actually no different from the controls. In rats depleted of serotonin using the synthesis blocker P-chlorophenylalanine (PCPA) or in rats in which the serotonergic terminals had degenerated following an injection of the neurotoxin 5,7-dihydroxytryptamine (5,7-DHT), the low dose of atropine was sufficient to produce a marked impairment of the spatial memory ability. A similar impairment of spatial memory ability was seen in rats with a combined lesion of the septum, the source of the cholinergic input to the hippocampus, and the serotonergic system using 5,7-DHT.

The presence of a raphe graft in the hippocampus protected the rats from the behavioral deficits evoked by combined damage to serotonergic neurons and blockade of cholinergic neurotransmission (Fig. 3). Thus, an injection of a low dose of atropine did not impair performance of raphe-grafted rats in the water maze (Richter-Levin and Segal, submitted).

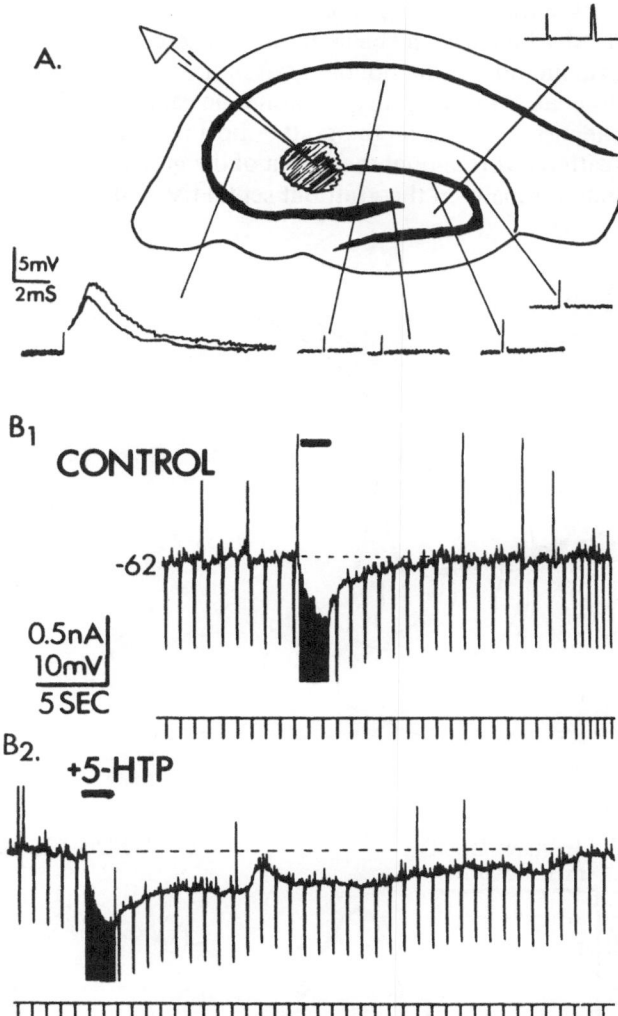

Fig. 1A, B. Graft–host interaction studied in an in vitro slice preparation. **A** Intracellular recording from a neuron in a graft drawn schematically as a *shaded* area inside the hippocampal slice. Selected regions of the hippocampal slice were stimulated with a monopolar microelectrode. While stimulation of most regions examined did not produce any clear postsynaptic effects, stimulation of the dentate hilus (*top right* specimen record) evoked an antidromic action potential discharge of the recorded neuron. This indicates that grafted neurons project axons into the dentate hilus where the normal serotonergic innervation is dense. The specimen records illustrate a stimulation artifact (on the *left*) and a response (on the *right*). Stimulation of region CA3 (*bottom left*) produced an EPSP in the grafted neuron indicating that CA3 neurons may branch and innervate the grafted cells. **B** The effects of a tetanic stimulation of the graft and membrane properties of a host hippocampal neuron. Initially in **B₁** a tetanic stimulation (*horizontal bar*) produces only a transient effect. After the addition of the serotonin precursor 5-HTP in **B₂**, which is taken up by serotonergic terminals, the stimulation produces a prolonged hyperpolarization associated with an increase in input conductance estimated by passage of constant hyperpolarizing current pulses (*lower records*). Scale in **A** is only for the lower records. Scale in **B₂** is the same as for **B₁**

We then tested the possible beneficial effect of a combined raphe/septal graft on physiological and behavioral parameters in the hippocampus. A morphological examination of the double graft indicated that serotonergic and cholinergic cells do coexist in the graft. Serotonergic cells detected with serotonin immunocyto-chemistry line up on the graft − host boundary whereas the cholinergic cells are scattered throughout the extent of the graft. Raphe tissue does not expand in volume normally in the graft but septal tissue does. A combined graft of raphe and

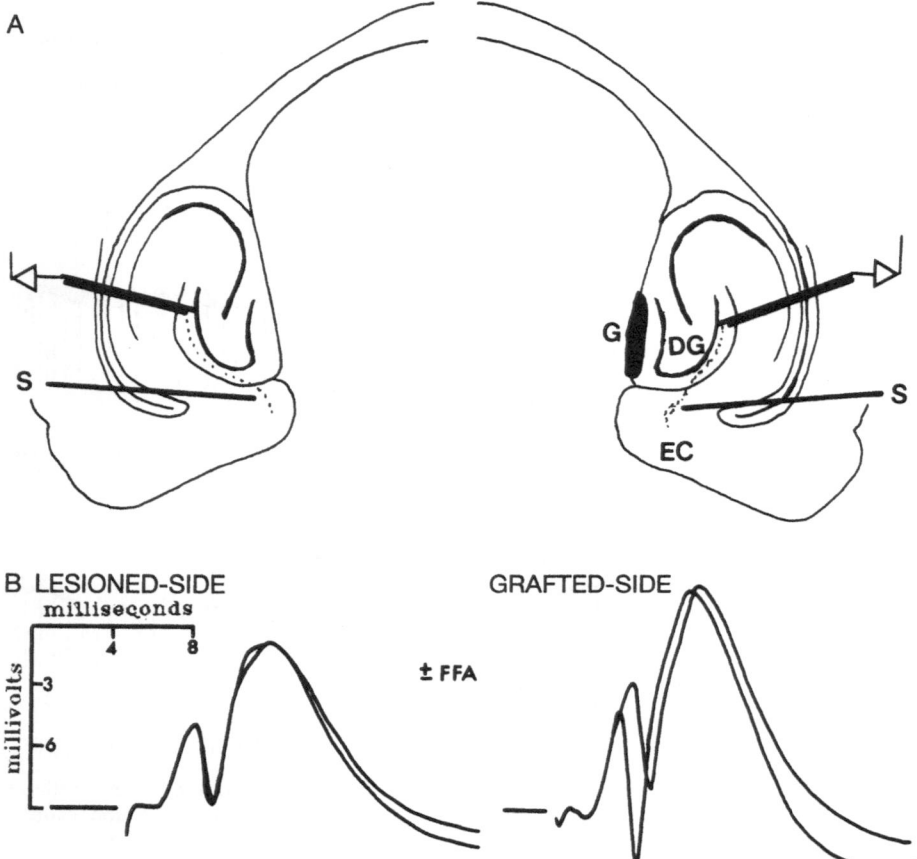

Fig. 2A, B. A Configuration of the stimulating and recording electrodes in a serotonin-depleted, unilateral raphe-grafted brain. Animals are anesthetized with chloral hydrate and placed in a stereotaxic head holder. Recording is made from the dentate gyrus *(DG)* while stimulating *(S)* the perforant path. The graft *(G)* is typically positioned on the medial wall of the dentate gyrus, lining the lateral ventricle. *EC,* entorhinal cortex. **B** Effects of a serotonin-releasing drug fenfluramine *(FFA)* on evoked response of the dentate gyrus to perforant path stimulation. The population response to perforant path stimulation was recorded in both hemispheres as seen in **A.** The drug was then injected intraperitoneally and the responses to stimulation recorded some 10−15 minutes later. The response in the lesioned side did not change much following drug application, whereas the response in the grafted hemisphere increased considerably following FFA. Such an increase can be seen in normal rat hippocampus. (From Segal et al. 1988)

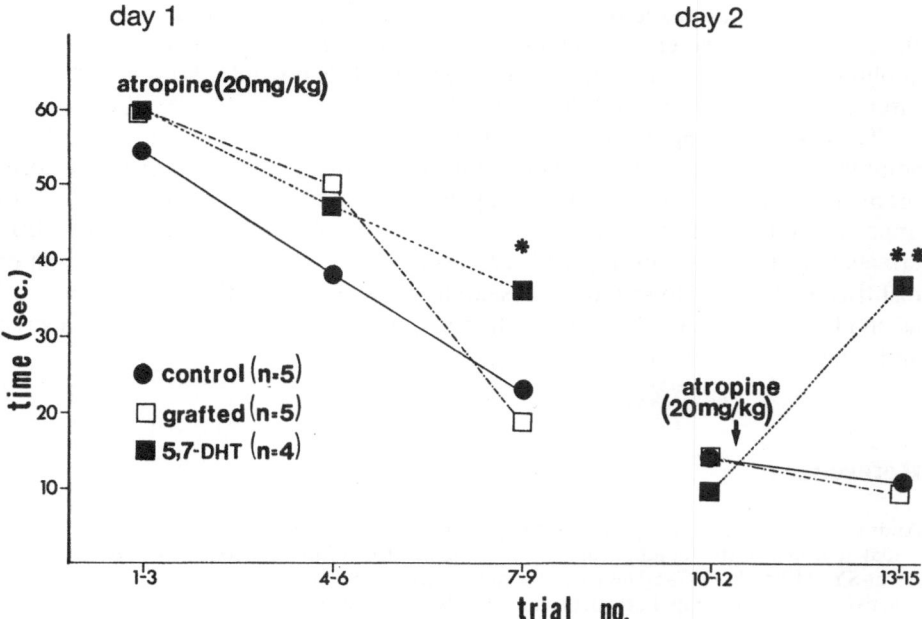

Fig. 3. Performance in a water maze is impaired in rats bearing a combined interruption of serotonin and ACh neurotransmission, and is repaired by a raphe graft. Three groups of rats were used: normal control, 5,7-DHT-treated rats in which brain serotonin was drastically reduced, and raphe-grafted rats which were similar to the 5,7-DHT-treated rats except that they were grafted with embryonic raphe cells. The experiment was conducted 2−3 months after grafting. Initial training was done in the presence of 20 mg/kg atropine injected to all rats prior to training. There were few differences among the groups except that at the end of the first day of training the lesioned group did not perform as well as the other groups. The first three trials of the second day of training were run without atropine. A low dose (20 mg/kg i.p.) of atropine was then injected. The drug markedly impaired the water maze performance only in the lesioned group, but not in the lesioned/grafted group. Ordinate, time to reach the submerged platform in the water maze. (From Richter-Levin and Segal 1988)

septum appears to retain properties of a septal graft and grow considerably in size in the host hippocampus. Physiologically, a combined graft appears to function as well as a raphe graft alone. Behaviorally, there is no obvious additive value to a combined raphe/septal graft. These are preliminary results and should be replicated in a more extensive study.

The results presented here indicate that there is an interaction between the cholinergic and serotonergic systems to affect cognitive functions associated with the hippocampus. The experiments with the raphe grafts in the hippocampus add an important dimension by suggesting that such an interaction takes place in the hippocampus proper and not in regions afferent to it. The nature of this interaction is yet unclear. It is possible that ACh enables the action of serotonin (or vice versa) or that each of them contributes an additive effect to the final decision point. At the physiological level studied in a slice preparation, we could not detect an effect of serotonin on the reactivity of hippocampal neurons to ACh, or vice versa. By the same token, we could not find an effect of serotonergic manipulations on cholinergic muscarinic receptors and vice versa. It is possible, however, that the interaction takes place at a different level.

These studies support the view that marked behavioral deficits can result from combined damage to two complementary neurotransmitter systems in the fore-brain and that a possible treatment may involve a simultaneous activation of both impaired systems. The graft as an experimental tool can be used to direct drug effects to a specific brain region. It can also serve as a source for replacement of lost tissue. Needless to say, more research is needed before the graft can be used as an effective treatment for AD patients, but its prospectives for both research and treatment are sound.

References

Andrade R, Nicoll RA (1987) Pharmacologically distinct actions of serotonin on single pyramidal neurones of the rat hippocampus recorded in vitro. J Physiol (Lond) 394:99−124

Assaf SY, Miller JJ (1978) The role of a raphe serotonin system in the control of septal unit activity and hippocampal desynchronization. Neuroscience 3:539−550

Barbeau A, Growdon JH, Wurtman RJ (eds) (1979) Choline and lecithin in brain disorders. Raven, New York

Bliss TVP, Dolphin AC (1982) Mechanisms of long term potentiation. Trends Neurosci 5:289−290

Colino A, Halliwell JV (1987) Differential modulation of three separate K conductances in hippocampal CA1 neurons by serotonin. Nature 328:73−77

Corkin S, Davis K, Growdon JH, Usidin E, Wurtman RJ (1982) Alzheimer's disease: A report of progress in research. Raven, New York

Davies SP, Maloney AJF (1976) Selective loss of central cholinergic neurons in Alzheimer's disease. Lancet II:1403

Deutsch JA (1983) The cholinergic synapse and the site of memory. In: Deutsch JA (ed) The physiological basis of memory. Academic, New York, pp 367−386

Dunnett SB, Low CW, Iversen SD, Stenevi U, Björklund A (1982) Septal transplants restore maze learning in rats with fornix-fimbria lesions. Brain Res 251:335−348

Freed WJ (1983) Functional brain tissue transplantation: Reversal of lesion-induced rotation by intraventricular substantia nigra and adrenal medulla grafts, with a note on intracranial retinal grafts. Biol Psychiatry 18:1205−1267

Morris R (1984) Developments of a water maze procedure for studying spatial learning in the rat. J Neurosci Methods 11:47−60

Perry EK, Tomlinson BE, Blessed G, Bergmann K, Gibson PH, Perry RH (1978) Correlation of cholinergic abnormalities with senile plaques and mental test scores in senile dementia. Br Med J 2:1457−1459

Richter-Levin G, Segal M (1988) Serotonin releasers modulate reactivity of the rat hippocampus to afferent stimulation. Neurosci Lett 94:173−176

Richter-Levin G, Segal M (1989) Raphe alls grafted into the hippocampus can ameliorate spatial memory deficits in rats with combined serotonergic/cholinergic deficiencies. Brain Res 478:184−186

Rossor M, Iversen LL (1986) Non-cholinergic neurotransmitter abnormalities in Alzheimer's disease. Br J Med Bull 42:70−74

Segal M (1975) Physiological and pharmacological evidence for a serotonergic projection to the hippocampus. Brain Res 94:115−131

Segal M (1980) The action of serotonin in the hippocampal slice preparation. J Physiol (Lond) 303:423−439

Segal M (1987) Interactions between grafted serotonin neurons and adult host rat hippocampus. Ann NY Acad Sci 495:285–295

Segal M, Azmitia EC (1986) Fetal raphe neurons grafted into the hippocampus develop normal adult physiological properties. Brain Res 363:162–166

Segal M, Azmitia E, Björklund A, Greenberger V, Richter-Levin G (1988) Physiology of graft-host interactions in the rat hippocampus. Prog Brain Res 78:95–102

Stenevi U, Björklund A, Svendgaard NA (1976) Transplantations of central and peripheral monoamine neurons to the adult rat brain: Techniques and conditions for survival. Brain Res 114:1–20

Strömberg I, Johnson S, Hoffer B, Olson L (1985) Reinnervation of dopamine-denervated striatum by substantia nigra transplants: immunohistochemical and electrophysiological correlates. Neuroscience 14:981–990

Vanderwolf CH (1987) Near-total loss of "learning" and "memory" as a result of combined cholinergic and serotonergic blockade in the rat. Behav Brain Res 23:43–57

Winson J (1980) Influence of raphe nuclei on neuronal transmission from perforant pathway through dentate gyrus. J Neurophysiol 44:937–950

Serotonin-Expressing Cells from Different Microregions of the Embryonic Rat Rhombencephalon: Behaviour in Cell Culture and in Transplants to the Adult Spinal Cord

N. König, N. Rajaofetra, M. J. Drian, F. Favier, F. Sandillon, C. Fuentès, and A. Privat

Summary

Physical properties of rat embryonal cells from different rhombencephalon regions were analysed using flow cytometry. Regional differences in forward-angle as well as right-angle scatter characteristics were detected. These differences might partly correspond to maturation gradients.

The behaviour of such cells in general, and of serotonin-expressing subpopulations in particular, was studied in cell culture and after transplantation into the completely transected adult rat spinal cord.

Rostral rhombencephalon cultures contained many more, and more intensely stained, immunoreactive cells than caudal rhombencephalon cultures, and particularly cultures from the caudal-most part of the rhombencephalon (containing the anlage of the B1–B2 groups). In contrast, caudal rhombencephalon tranplants derived from the same cell suspensions did very well in the spinal cord. There, the number of immunoreactive cells was at least equivalent to that of rostral rhombencephalon transplants.

The patterns of neurite outgrowth into the host spinal cord were conditioned by general environmental factors (e.g., the position of the grey versus white matter interface, or the proximity of the central canal) but they also depended upon the microregional origin of the transplanted cells. Since these different innervation patterns presumably correspond to different types and degrees of functional restoration, it probably is useful to select cells for grafting not only with respect to their neurotransmitter but also with respect to other criteria, including their microregional origin.

Introduction

Transplantation is increasingly used to restore functions in damaged brains, either for experimental or therapeutical purposes. The choice of the cells to be transplanted usually has been determined according to one of the following strategies:
1. The *"minipump strategy"* in which grafted cells were simply intended to produce a given substance and to release it into the host tissue, e.g., chromaffin tis-

sue from the adrenals has been used to provide the brains of Parkinson patients with catecholamines (Backlund et al. 1985).
2. The *"network strategy"* in which the grafted cells were intended to establish appropriate and functional synaptic connections with the host brain, e.g., substantia nigra neurons have been transplanted into the denervated striatum (Björklund and Stenevi 1979; Mahalik et al. 1985).

Both approaches have their advantages and their disadvantages. The minipump strategy allows a larger choice of the possible candidates for grafting, including cells located outside the central nervous system as in the example mentioned above. The possibility of making autografts is most welcome for reducing the risks of immunological rejection. In many cases, however, the network strategy seems to provide better results, in terms of long-term survival of the grafted cells as well as restoration of functions.

Whenever the network strategy is the best choice, the appropriate selection of the cell material for grafting is essential. In the present paper, we specifically address the behaviour of embryonic serotonin-expressing cells that were either grown in vitro or transplanted into the adult spinal cord (Privat et al. 1986a, b, 1988; König et al. 1987b, 1988a). However, the principal conclusion, stressing the importance of the microregional origin of the transplanted cells, may be pertinent for nerve cell grafting in general.

Materials and Methods

Sprague–Dawley rat embryos (Iffa Credo) were dissected on embryonal days (ED) 13, 14, and 15, the day following mating being considered as day 0. The mean crown–rump lengths at these stages were 8.5, 10.5, and 12 mm, respectively. We used immunostained whole mounts showing the position of the monoaminergic cell groups in the embryonic brainstem (König et al. 1988b, c) as guides for the microdissection of specific rhombencephalon regions. The rhombencephalon was cut into two pieces:
1. a caudal piece (containing the anlage of the B1–B3 groups) extending from the cervical end of the spinal cord to the pontine flexure, and
2. a rostral piece (containing the anlage of B4–B9 complex) extending from the pontine flexure to the rhombencephalic isthmus.

We will refer to the two pieces as "caudal rhombencephalon" and "rostral rhombencephalon", respectively. Notice that the latter piece includes the raphe groups often designated by the misnomer "mesencephalic raphe" (for discussion, see König et al. 1988b). For some experiments, the rostral rhombencephalon was further subdivided into its alar-plate and its basal-plate parts, and the caudal rhombencephalon was subdivided into a part containing B1–B2 and another containing B3.

The cells were mechanically dissociated in Puck's Saline A (Gibco) without enzymatic digestion. They then were spun down at $100 \times g$ for 10 min and resuspended either in Puck's Saline for flow cytometric analysis or in culture medium for transplantation or cell culture experiments. The culture medium contained 5% Nuserum (Collaborative Research).

The forward-angle scatter and the right-angle scatter properties were determined using an ORTHO 50H flow cytometer equipped with an argon ion laser (488 nm) and ODAM-MCA 3 000 computer.

For the culture experiments, the cells were plated on poly-D-lysine (Boehringer) in 24-well plates. The standard plating density was 4 000 cells per mm^2. However, we also experimented with plating densities as low as 200 cells per mm^2. The cells were cultured at 35 °C, in a humidified atmosphere containing 5% CO_2. After 4−21 days in vitro, the cells were fixed with glutaraldehyde (2.5%) and processed for the immunocytochemical detection of serotonin using a polyclonal antibody (Privat et al. 1986b).

The density of immunostaining was quantified using a CCD camera (Sony) and a Jasmin computer (Tran) equipped with an image treatment and analysis board (Imaging Technology).

For the transplantation experiments, the spinal cord of adult rats was completely transected at the lower thoracic level. One week later, either culture medium alone or 3−5 µl of a concentrated cell suspension (30 000−50 000 cells per µl) was injected below the transection site. After delays ranging from one week to several months, the host animals were pretreated with pargyline (100 mg/kg) and L-tryptophan (100 mg/kg), perfused with glutaraldehyde (5%), and processed for the immunocytochemical detection of serotonin (Privat et al. 1986b).

Results

Regional Differences in General Physical Cell Properties

The flow cytometric analysis showed that cells from the ED 14 caudal rhombencephalon had higher right angle scatter values than cells from the rostral rhombencephalon. Marked regional differences in right-angle scatter as well as forward-angle scatter were still observed in ED 15 cell suspensions: there was a rostro-caudal increase in right-angle scatter values (Fig. 1a), and the alar-plate population of the rostral rhombencephalon differed from the other populations by the presence of a sharp forward-angle scatter peak (Fig. 1b).

Serotonin-like Immunoreactivity in Cultured Cells Derived from Different Rhombencephalon Regions

Cells expressing serotonin (dilution of the primary antibody 1:25 000) were found in cultures of caudal as well as rostral rhombencephalon. However, the number of the detected cells and the intensity of immunostaining drastically varied as a function of the regional origin of the cells. Cultures prepared from the rostral rhombencephalon (including the anlage of the B4−B9 complex) contained numerous immunoreactive cells, the majority of which were intensely stained (Fig. 2a). In cultures from the region just caudal to the pontine flexure (including B3) fewer

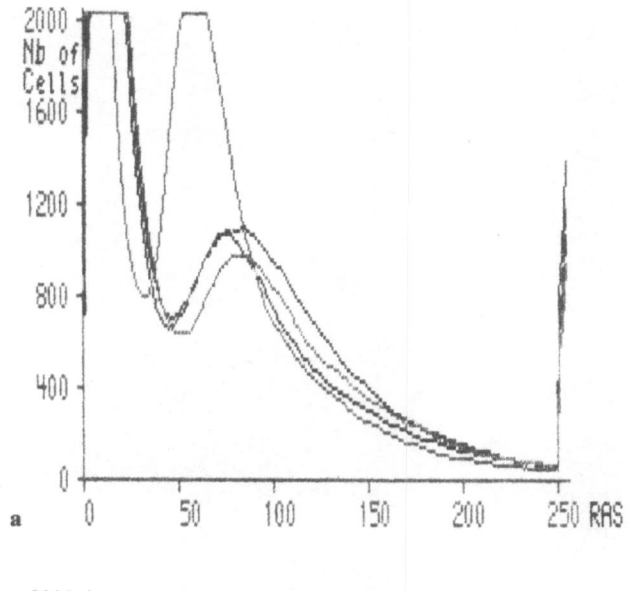

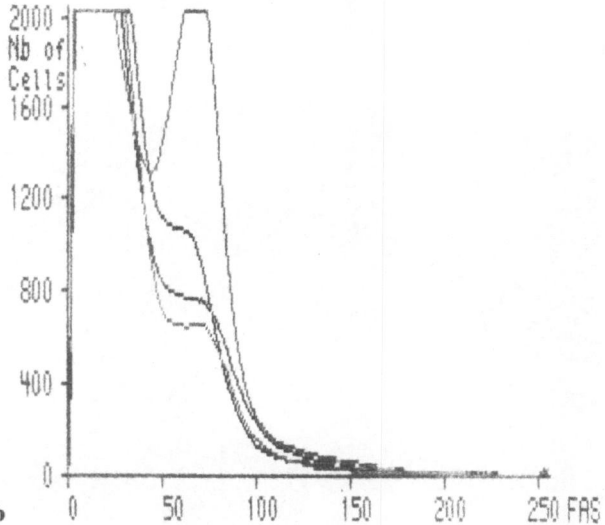

Fig. 1a, b. a Right-angle scatter *(RAS)* and **b** forward-angle scatter *(FAS)* histograms of dissociated ED 15 rhombencephalon cells. *Green:* alar-plate part of the rostral rhombencephalon; *red:* basal-plate part of the rostral rhombencephalon (including the anlage of the B4−B9 complex); *black:* rhombencephalon-part situated just caudal to the pontine flexure (including B3); *blue:* caudal-most rhombencephalon-part (including B1−B2)

cells were immunoreactive. Although some of them were strongly labelled (Fig. 2b), many were only moderately (Fig. 2c) or weakly labelled. As to the cultures prepared from the caudal-most rhombencephalon pieces (including B1−B2), very few immunoreactive cells were detectable and most of them were weakly labelled (Fig. 2d). Although the number of detectable B1−B2 cells increased when the primary antibody was diluted 1:15 000, it was still about 20 times lower than the number of immunoreactive cells in rostral rhombencephalon cultures.

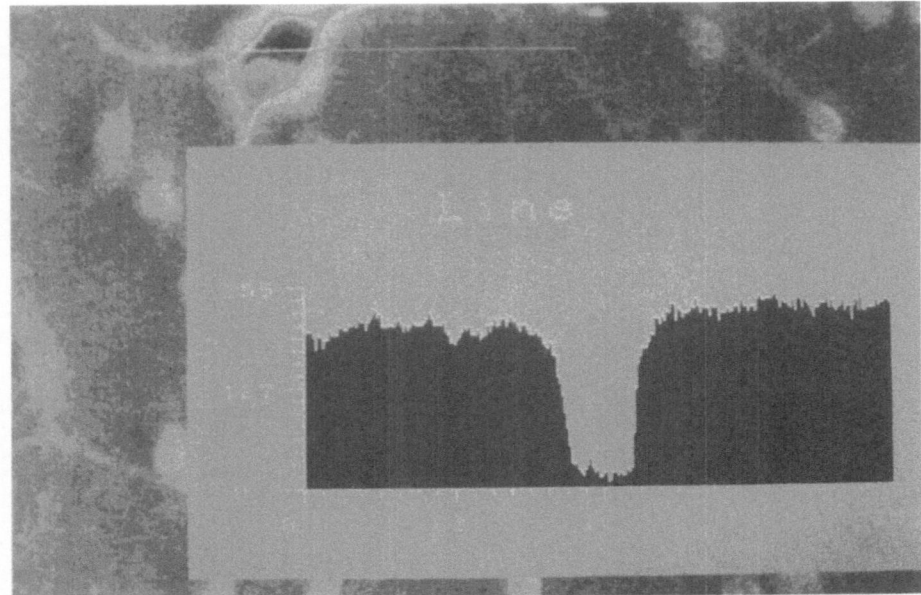

a

b

Fig. 2. Pseudocolour translations of the intensity of serotonin-like immunoreactivity (*blue, background labelling; green,* weak labelling; *yellow,* medium labelling; *red,* strong labelling) and histograms of the optical density along the light blue horizontal lines across the cell bodies. The cells shown are from cultures containing **a** B4−B9, **b, c** B3, and **d** B1−B2 populations

Fig. 2c

Fig. 2d

Growth and "Social Behaviour" of Cultured Cells Derived from Different Rhombencephalon Regions

In caudal rhombencephalon cultures, serotonin-expressing cell bodies were almost exclusively found in close contact with other (generally nonimmunoreactive) cells. Their neurites seemed to closely follow cell or process alignments in their neigbourhood (Fig. 3a). Conversely, many rostral rhombencephalon cells were found in relatively isolated positions and their neurites seemed to be less dependent upon cell-cell contact (Fig. 3b). These neurites often spanned across cell-free spaces in a more or less rectilinear way, in contrast with the meandering course of caudal-cell neurites.

Behaviour of Cells Taken from Different Rhombencephalon Regions After Transplantation into the Adult Spinal Cord

Dissociated embryonic cells transplanted into the completely transected adult spinal cord were able to express serotonin and to survive for several months. The transplanted cells usually formed nucleus-like clusters which could be easily distinguished from the isolated intrinsic serotonic cells.

In contrast to the relative quantities of immunoreactive cells observed in the cultures, the number of serotonin-expressing cells in caudal rhombencephalon grafts was at least equivalent to that in rostral rhombencephalon grafts. Grafts of the rhombencephalon part containing the B3 group (Fig. 4) and even grafts of the caudal-most rhombencephalon comprising B1−B2 (Fig. 5) produced numerous and well-differentiated immunoreactive neurons.

The innervation pattern produced by the grafted neurons was manifestly influenced by environmental factors. For instance, the interface between grey and white matter seemed to constitute an obstacle for the progression of outgrowing fibres (in cases where embryonic cells had been deposited in both the grey and white matter, outgrowing neurites were seen coursing on either side of the interface, but rarely crossing it). Conversely, the area surrounding the central canal was a propitious territory for growing neurites, irrespective of the regional origin of the grafted cells (Fig. 6a). Another generally well innervated territory was the intermediolateral area, although the most striking innervation was seen in animals that had received B1−B2 cells (Fig. 5). As to the dorsal and ventral horns, their innervation seemed to depend to a greater extent upon the microregional origin of the grafted cells. Serotonin-expressing cells from the whole rostral rhombencephalon seemed to innervate the ventral horns slightly more densely than the dorsal horns, and this tendency was increased when cell suspensions that did not contain the whole embryonic B4−B9 complex, but only its ventro-lateral part, were grafted (Fig. 6b, c). Conversely, grafts comprising the embryonic B3 group produced a very dense innervation of the superficial dorsal horn layers (Fig. 6d). The described preferential innervation patterns were usually more clear-cut at some distance from the graft than close to it.

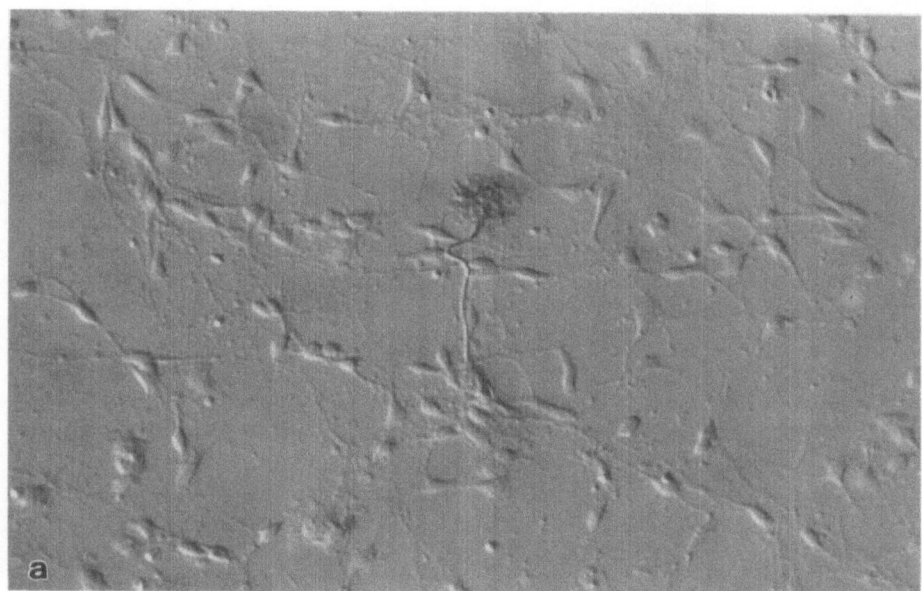

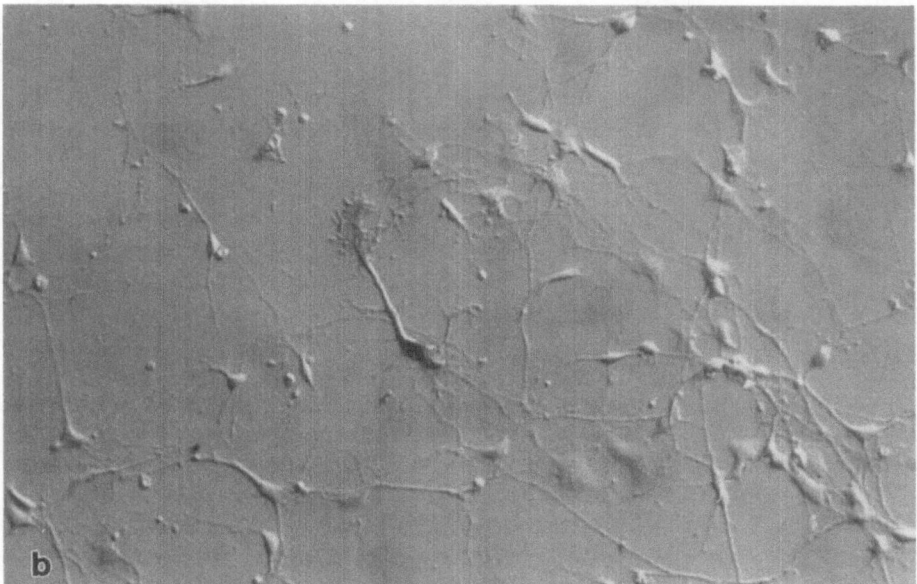

Fig. 3. Differential-interferential contrast micrographs of serotonin-expressing cells in **a** caudal rhombencephalon culture and **b** rostral rhombencephalon culture. Notice the difference in "social behaviour". × 275

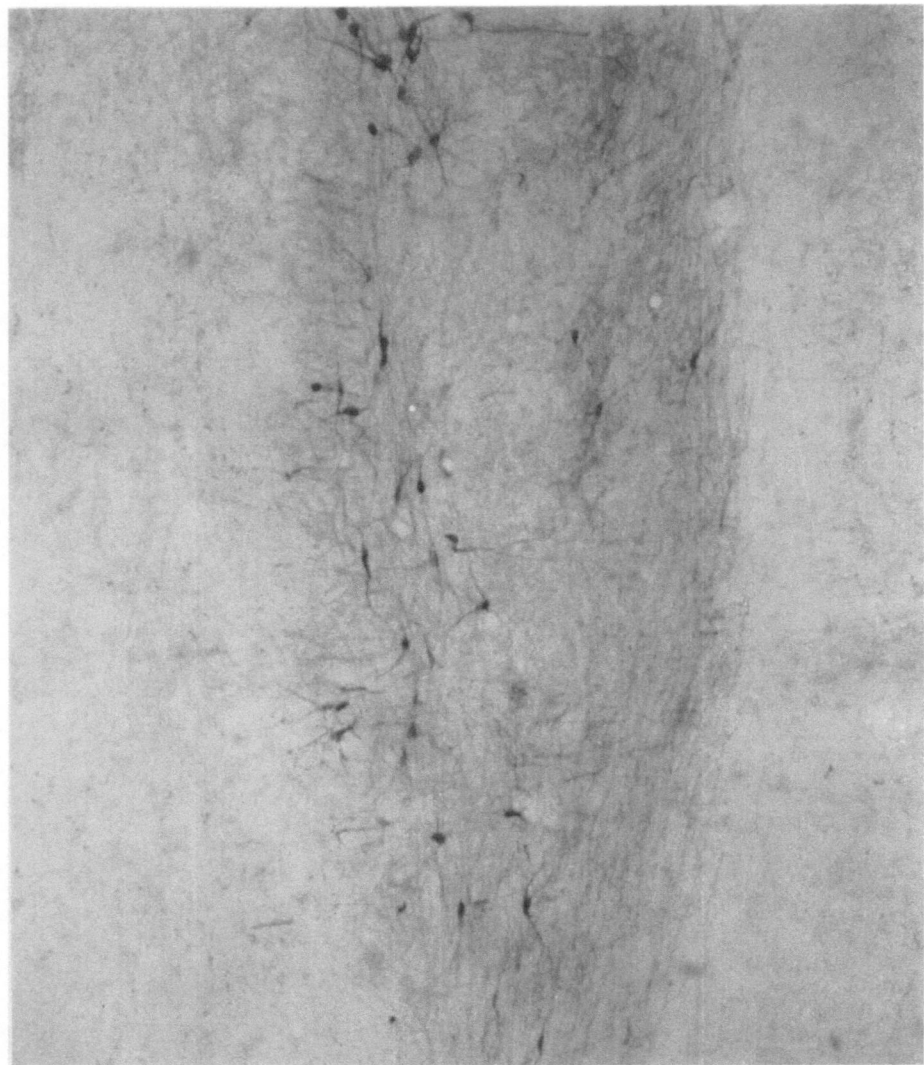

Fig. 4. Micrograph of a horizontal spinal cord section, showing serotonin-expressing B3 cells transplanted below the transection site. Rostral is up. Notice that many cell processes are oriented perpendicularly to each other, which is similar to their normal orientation in the brainstem. × 100

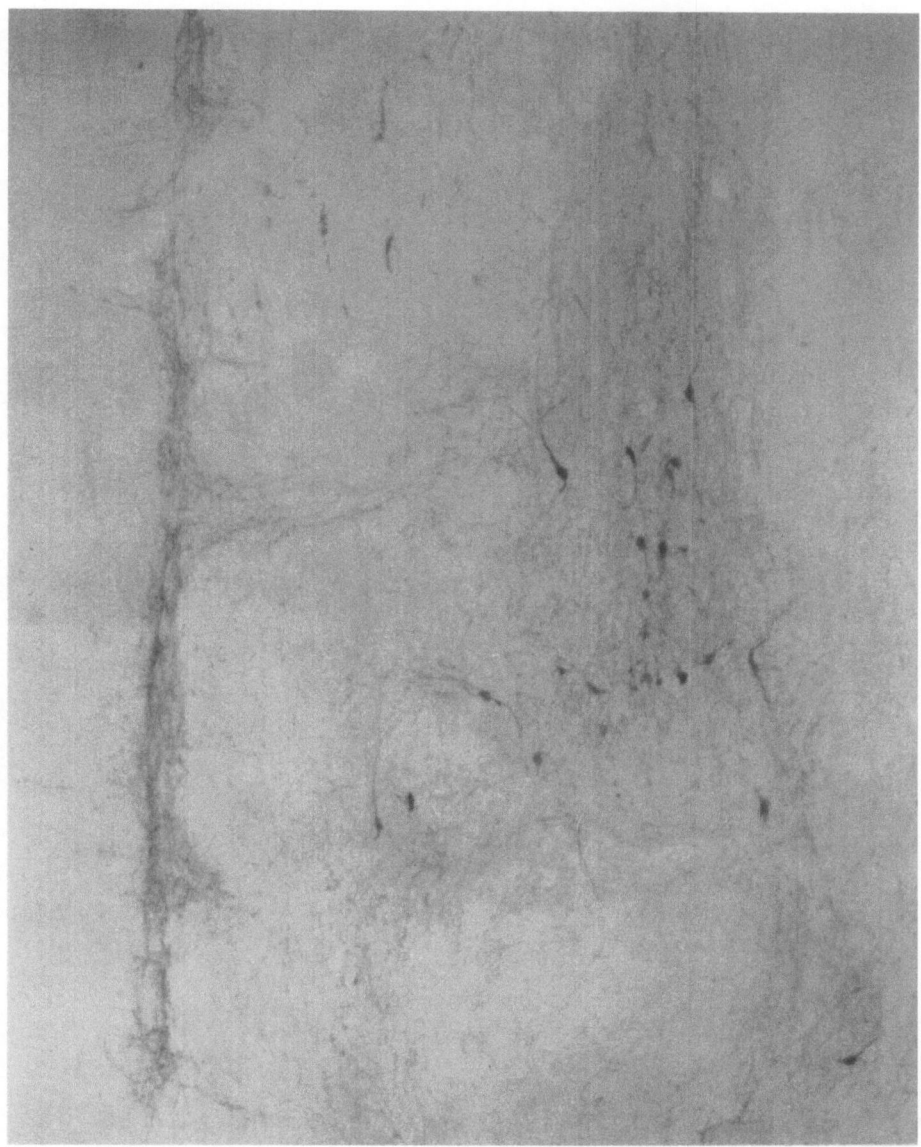

Fig. 5. Micrograph of a horizontal spinal cord section, showing serotonin-expressing B1–B2 cells transplanted below the transection site. Rostral is up. Notice the intense and sharply delineated innervation of the intermediolateral column (at *left*). × 100

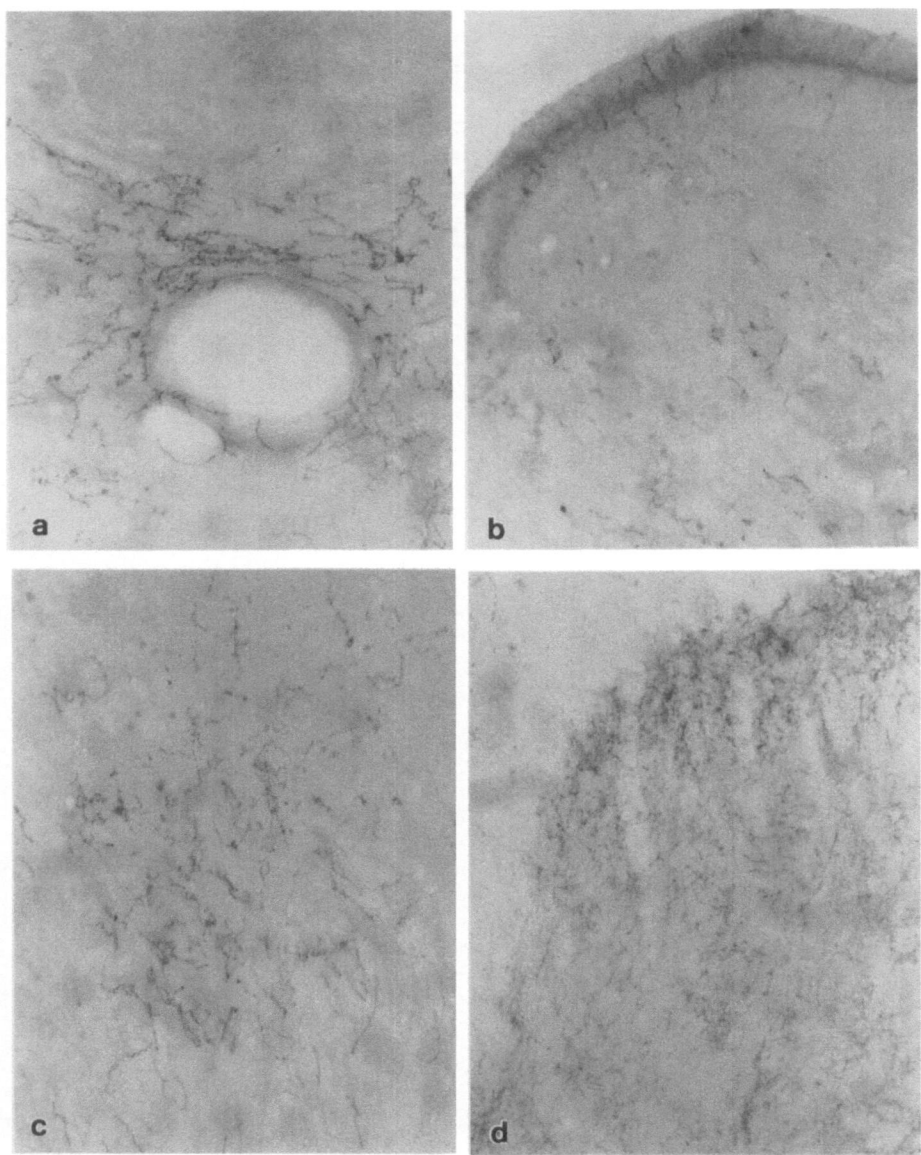

Fig. 6a–d. Micrographs of transversal spinal cord sections, showing neurites arising from serotonin-expressing cells transplanted below the transection site. **a** Neurites close to the central canal; the graft contained cells from the ventrolateral part of the embryonic B4–B9 complex. **b** Same graft, very sparse dorsal horn innervation. **c** Same graft, ventral horn; the innervation is denser than in the dorsal horn shown in **b. d** B3 graft, dorsal horn; very dense innervation of the superficial layers, similar to the normal pattern. × 185

Discussion

The flow cytometric analysis has shown that cell populations from different rhombencephalon regions significantly differ in physical properties such as size or density. The rostro-caudal increase in right-angle scatter values might correspond to a maturation gradient since the neurogenesis in the rat caudal rhombencephalon has been reported to start one or two days earlier than in the rostral rhombencephalon (Altman and Bayer 1981). At ED 15, a sharp forward-angle scatter peak was present only in the alar-plate part of the rostral rhombencephalon (which contains the rapidly growing cerebellar anlage); this suggests that the cell proliferation in the remaining rhombencephalon regions was slowing down at that stage. Since the expression of serotonin in the rat caudal rhombencephalon starts about two days later than in the rostral rhombencephalon (Olson and Seiger 1972; Wallace and Lauder 1983; König et al. 1988b), there may exist considerable regional differences in the delay between the last mitosis on the one hand, and the expression of detectable amounts of serotonin on the other. It would be interesting to know if there is any link with the fact that many caudal serotonergic neurons coexpress substance P and/or thyrotropin-releasing hormone (TRH) (Johansson et al. 1981; Foster et al. 1985b; Towle et al. 1986).

Such regional differences may be expression of population-specific genetic programmes; alternatively, the delayed onset of serotonin-like immunoreactivity observed in the caudal populations might be due to epigenetic factors such as reduced autostimulation (De Vitry et al. 1986) or inhibitory effects exerted by the serotonergic cells themselves (Haydon et al. 1984; see also König et al. 1987a) or by their neighbours.

In our culture experiments with populations derived from rhombencephalic microregions we observed marked rostro-caudal differences in the number of immunoreactive cells. Although the number of serotonergic neurons in vivo is lower in the caudal than in the rostral rhombencephalon, the estimated proportion (approximately 1:2.5; see Foster et al. 1985b) is insufficient to fully explain the paucity of immunoreactive cells observed in our caudal rhombencephalon cultures. Although these results may indicate regional genetic differences, the above-mentioned epigenetic factors cannot be ruled out, since the cultures did not only contain potentially serotonin-expressing cells of a given population, but also many other neuronal and nonneuronal cells.

Another striking difference related to the microregional origin of the cultured cells was the close spatial association of caudal serotonin-expressing cells with their neighbours, as opposed to the more independent growth of rostral cells. This difference is apparently not limited to serotonergic neurons nor to neurons in general, since similar differences have been reported for glial cell cultures derived from caudal versus rostral rhombencephalon (Wilkie and Lauder 1988). The mechanisms underlying our morphological observations probably concern regional differences in adhesion molecule activity, which is an important factor in the regulation of neurite outgrowth.

The interplay between environmental factors on the one hand, and the regional origin of the cells on the other, was particularly obvious in our grafting experiments. The completely transected adult spinal cord provided a great number of environmental constraints and clues ranging from general ones, such as the position of the grey versus white matter interface or the proximity of the central canal, to more specific ones, such as the vacancy of specific synaptic sites in particular layers or regions. The behaviour of serotonin-expressing embryonic cells transplanted into this environment manifestly depended upon their microregional origin. Cells taken from a given rhombencephalon region preferentially (but not exclusively) innervated a given layer or group of layers. The fact that these preferences grossly corresponded to the normal innervation patterns of the spinal cord (Steinbusch 1984) raises the question whether the growth cones of transplanted embryonic neurons can be specifically guided towards their normal, but adult, target areas. In any case, it has been shown that grafted serotonergic cells are able to establish normal-looking and presumably functional synapses with spinal neurons (Privat et al. 1986b, 1988).

Another parameter that seems to be influenced by the microregional origin is the proportion of grafted cells that are able to survive and to express detectable amounts of serotonin in the spinal cord. As mentioned above, the proportion of serotonergic neurons in the caudal rhombencephalon is lower than that in the rostral rhombencephalon. When grafting the same total amount of cells (serotonergic plus other cells) one therefore should expect to find a lower number of serotonin-expressing cells in grafts of caudal rhombencephalon than in grafts of rostral rhombencephalon. However, we found at least equivalent, if not higher, numbers of immunoreactive cells in caudal grafts. This is in agreement with findings reported by Foster et al. (1985b), who transplanted "mesencephalic raphe" cells (corresponding to our "rostral rhombencephalon" cells) and "medullary raphe" cells (probably corresponding to B1−B2 cells) into the spinal cord after treatment with 5,7-dihydroxytryptamine. One possible explanation for the discrepancy between the expected and the observed numbers of immunoreactive cells is that caudal rhombencephalon cells might survive better since the spinal cord is the normal target area for most of them (in contrast with rostral rhombencephalon cells). Alternatively, the transplantation might either raise the serotonin levels in caudal cells that would have been undetected in the normal brain or lower the serotonin level in rostral cells below detection threshold. In any case, the caudal rhombencephalon cells seem to be favoured in the adult spinal cord environment. Conversely, rostral rhombencephalon cells seem to do better than caudal rhombencephalon cells when transplanted into the striatum (Foster et al. 1985a).

Conclusion

The behaviour of embryonic neurons is not only determined by the neurotransmitter they express, but also by other properties, some of which seem to be related to their microregional origin. When transplanted, cells of different regional origin produce markedly different innervation patterns and, presumably, different types and degrees of functional restoration.

When trying to attenuate the effects of lesions or degenerative processes it may be useful to be aware of these properties in order to make the best choice for a given problem. This does not necessarily mean that one should take the cell populations that normally innervate the lesioned area, although this might be a good option.

Acknowledgments. We would like to acknowlegde the photographical assistance of Mr. Jean-René Teilhac. This research was supported by grants from DRET, IRME, CNRS, INSERM, and USTL.

References

Altman J, Bayer SA (1981) Development of the brainstem in the rat. V. Thymidine-radiographic study of the time of origin of neurons in the midbrain tegmentum. J. Comp Neurol 198:677–716

Backlund EO, Granberg PO, Hamberger B, Knutson E, Martensson A, Sedvall G, Seiger A, Olson L (1985) Transplantation of adrenal medullary tissue to striatum in parkinsonism. J Neurosurg 62:169–173

Björklund A, Stenevi U (1979) Reconstruction of the nigrostriatal dopamine pathway by intracerebral nigral transplants. Brain Res 177:555–560

De Vitry F, Hamon M, Catelon J, Dubois M, Thibault J (1986) Serotonin initiates and autoamplifies its own synthesis during mouse central nervous system development. Proc Natl Acad Sci USA 83:8629–8633

Foster GA, Schultzberg M, Björklund A, Gage FH, Hökfelt T (1985a) Fate of embryonic mesencephalic and medullary raphe neurones transplanted to the striatum, hippocampus and spinal cord of the adult rat: analysis of 5-hydroxytryptamine, substance P and thyrotropin releasing hormone-immunoreactive cells. In: Björklund A, Stenevi U (eds) Transplantation in the mammalian CNS. Elsevier, Amsterdam, pp 179–189

Foster GA, Schultzberg M, Gage FH, Björklund A, Hökfelt T, Nornes H, Cuello AC, et al. (1985b) Transmitter expression and morphological development of embryonic medullary and mesencephalic raphe neurones after transplantation to the adult rat central nervous system. I. Grafts to the spinal cord. Exp Brain Res 60:427–444

Haydon PG, McCobb DP, Kater SB (1984) Serotonin selectively inhibits growth cone motility and synaptogenesis of specific identified neurons. Science 226:561–564

Johansson O, Hökfelt T, Pernow B, Jeffcoate SL, White N, Steinbusch HWM, Verhofstad AAJ, et al. (1981) Immunohistochemical support for three putative transmitters in one neuron: Coexistence of 5-hydroxytryptamine, substance P, and thyrotropin releasing hormone-like immunoreactivity in medullary neurons projecting to the spinal cord. Neuroscience 6:1857–1881

König N, Han VKM, Lieth E, Lauder J (1987a) Effects of coculture on the morphology of the identified raphé and substantia nigra neurons from the embryonic rat brain. J Neurosci Res 17:349–360

König N, Mansour H, Drian MJ, Sandillon F, Favier F, Marlier L, Privat A (1987b) Caudal versus rostral regions of embryonic rat rhombencephalon: Flow cytometry, cell culture, immunocytochemistry, and transplantation. Soc Neurosci 13:570

König N, Rajaofetra N, Drian MJ, Sandillon F, Fuentès C, Favier F, Privat A (1988a) Differentiation and growth of embryonic serotonergic cells in culture and in transplants to the adult rat spinal cord: genetic and epigenetic factors (Abstr) Int J Dev Neurosci [Suppl 1] 6:43

König N, Wilkie MB, Lauder J (1988b) Tyrosine hydroxylase and serotonin containing cells in embryonic rat rhombencephalon: a whole-mount immunocytochemical study. J Neurosci Res 20:212–223

König N, Wilkie MB, Lauder J (1989) Dissection of monoaminergic neuronal groups from embryonic rat brain. In: Shahar A, Haber B (eds) A dissection and tissue culture manual of the nervous system. Liss, New York (in press)

Mahalik TJ, Finger TE, Strömberg I, Olson L (1985) Substantia nigra transplants into denervated striatum of the rat: ultrastructure of graft and host interconnections. J Comp Neurol 240:60–70

Olson L, Seiger A (1972) Early prenatal ontogeny of central monoamine neurons in the rat: fluorescence histochemical observations. Z Anat Entwicklungsgesch 137:301−316

Privat A, Mansour H, Geffard M, Lerner-Natoli M (1986a) Transplantation of 5-HT neurons to the adult brain. In: Briley M, Kato A, Weber M (eds) New concepts in Alzheimer's disease. Macmillan, London, pp 280−299

Privat A, Mansour H, Pavy A, Geffard M, Sandillon F (1986b) Transplantation of dissociated foetal serotonin neurons into the transected spinal cord of adult rats. Neurosci Lett 66:61−66

Privat A, Mansour H, Geffard M (1988) Transplantation of fetal serotonin neurons into the transected spinal cord of adult rats: morphological development and functional influence. Prog Brain Res 78:155−166

Steinbusch HWM (1984) Serotonin-immunoreactive neurons and their projections in the C.N.S. In: Björklund A, Hökfelt T, Kuhar MJ (eds) Handbook of chemical neuroanatomy, vol 3. Elsevier, Amsterdam, pp 68−125

Towle AC, Breese GR, Mueller RA, Hunt R, Lauder JM (1986) Early postnatal administration of 5,7-dihydroxytryptamine: Effects on substance P and thyrotropin-releasing hormone reurons and terminals in rat brain. Brain Res 363:38−46

Wallace JA, Lauder JM (1983) Development of the serotonergic system in the rat embryo: an immunocytochemical study. Brain Res Bull 10:459−479

Wilkie MB, Lauder JM (1988) Radial glial cultures from embryonic rat brain: regional heterogeneity and the radial glial lineage (Abstr). Annu Meet Eur Neurosci Assoc 11:239

Two Approaches for the Reversal of Phenobarbital-Induced Behavioral Birth Defects

J. Yanai, F. Fares, M. Gavish, Z. Greenfeld, C. G. Pick,
Y. Rogel-Fuchs, and D. Trombkal

Summary

The present chapter presents a model for the reversal of behavioral birth defects. Since the neuroteratogen employed, phenobarbital, alters numerous processes and behaviors, the study focused on alterations in the hippocampus and its related behaviors. Mice were exposed to phenobarbital prenatally, although some of the experiments also included neonatally exposed groups. At adulthood they showed deficits in spontaneous alternation, Morris maze and eight-arm maze tests. Studies on hippocampal morphology revealed areal and cell losses and deficient dendritic architecture in the surviving neurons, including reductions from control in the number of dendritic branches, area, and spine density, but wider fission angle than control. Neurochemical studies on the hippocampus revealed the following alterations:
1. decrease in norepinephrine (NE) level and the number of NE cell bodies,
2. no change in the serotonergic system,
3. an increase in muscarinic receptors B_{max} in the hippocampus,
4. transient decrease in gamma-aminobutyric acid (GABA) uptake, and an increase in the B_{max} of GABA and benzodiazepine receptors.

The changes in GABA did not correspond with the sensitive periods for the behavioral deficits. Transplantation of cholinergic neurons into the hippocampus of the treated mice reversed most of the deficits in eight-arm maze behavior, while transplantation of noradrenergic cells did not have an effect. Consistently, destruction of the inhibiting dopaminergic innervations in the septum increased hippocampal choline acetyltransferase (ChAT) activity and enabled the treated mice to reach normal performance in the maze after training.

Introduction

Brain tissue transplantation can now be successfully applied to alter a whole host of behaviors and functions or to correct them when they are deficient. However, correcting neural and behavioral birth defects induced by insults has rarely been attempted and usually with very limited success. The same difficulty has been experienced when the reversal of neural and behavioral birth defects has been attempted by other means. The paucity of application appears to be related to a lack of models rather than to feasibility. That is, understanding of the mechanism of a behavior or function is a prerequisite for its successful alteration by transplantation or by other manipulations. On the other hand, reversal can be easily done in experimental lesions (Perlow et al. 1979) where, obviously, the cause for the change is known. Even certain functional or behavioral states with genetic etiology or environmental causes can be altered (Werner and Wong 1987) when the mechanisms are fairly well understood. This explains why Parkinson's disease, whose mechanisms have been extensively studied, is the first target in the attempt to reverse neural defects with transplantation (Backlund et al. 1985).

On the other hand, the effect of most neuroteragotens is general, that is, they perturb numerous brain processes resulting in multiple behavioral alterations. Our investigations on phenobarbital neuroteratogenicity have been complicated by the fact that phenobarbital diffuses readily into all parts of the brain and its effect is rather general. We attempted to counter this methodological obstacle by studying, as end points, behaviors which are known to be primarily related to a certain brain region and specific innervations, while still keeping in mind that other brain regions and innervations are also affected by the drug. Among them, there may be structures which exert some control on the behaviors under study. It was hoped that this approach would permit identification of the mechanisms of the behavioral deficits. Studying behaviors which are mostly specific to a certain region or biochemical process makes it possible to attempt to reverse the deficits by experimental manipulations, which also serve as a probe to identify or confirm the mechanism of the deficits.

The present chapter presents a model of neuroteratogenicity. Although most of the research effort was devoted to prenatal exposure, some of the experiments included a neonatally exposed group, as the first neonatal week in mice is analogous to late pregnancy in humans. Pregnant female mice were exposed to phenobarbital during pregnancy. Their offspring were tested at adulthood for performance in the eight-arm maze (Olton and Samuelson 1976), Morris maze (Morris 1984), and spontaneous alternations (Roberts et al. 1962); all of these behaviors are believed to be largely dependent on the integrity of the septohippocampal cholinergic innervations. Parallel studies were conducted on the major innervations to the hippocampus. Finally, brain tissue transplantation and other manipulations of the hippocampal innervations were applied for the following reasons:
1. to establish a procedure to reverse the behavioral deficits and
2. as experimental probes which in concert with biochemical studies, facilitate understanding of the mechanism of the drug effect on the development of the behaviors in question.

An Animal Model for Phenobarbital Neuroteratogenicity

Early Exposure to Phenobarbital

Adult HS/I bg mice used as parents were housed in mating groups of one male and four females. Their offspring (the subjects of these experiments) received phenobarbital prenatally via the placenta. Some studies employed additional groups which were exposed to phenobarbital neonatally via daily injections (Yanai et al. 1979; Yanai and Bergman 1981).

Prenatal Administration. Female parent mice were checked every morning and those that conceived, as evidenced by the existence of a vaginal plug, were housed with other pregnant females. On gestation day 9 (the day in which the plug appeared was considered gestation day 1) the female mice were housed in individual cages. Treated female mice then received milled mouse food containing 3 g/kg phenobarbital in acid form (their only food source) and water, both available ad libitum. Control female mice received milled food and water. Drug administration continued until day 18, when the phenobarbital and control diets were replaced with regular mouse pellets.

The barbiturate-exposed and control offspring were maintained with their mothers under standard laboratory conditions and received no further drug treatment.

Neonatal Administration. In this group, parent female mice were not checked for vaginal plugs. Instead, they remained with the male mice until pregnancy became apparent, when they were moved to individual cages. After delivery, each litter was divided into control and barbiturate-treated groups. Toe clippings were used for identification. Treated pups received a daily subcutaneous injection of either 50 mg/kg or 40 mg/kg sodium phenobarbital in sterilized water (10 ml vehicle/kg mouse) on days 2 to 21 (delivery day = day 1). Control pups received vehicle injections. All pups were maintained after 21 days with their mothers under standard laboratory conditions with no further drug treatment.

Prenatally and neonatally treated pups, which were not used earlier for testing, as well as their respective controls, were weaned on day 23. They were segregated according to gender and maintained in groups of five under standard conditions until they were used for the various studies.

Behavioral Deficits

As discussed above, our approach was to study behaviors which may be linked, to a large degree, to the hippocampus. Several behaviors were studied and all were shown to be impaired by early phenobarbital exposure (see below). Among them, the eight-arm maze performance was chosen for further studies on the reversal of behavioral birth defects by experimental manipulations. This behavior is often studied in animals (Olton and Samuelson 1976) and even in humans (Aadland et al. 1985) as an indicator of spatial memory and spatial discrimination, and is thought to be largely related to the hippocampus (Olton et al. 1978) and, more specifically, to the septohippocampal cholinergic pathways. It has been demonstrated that specific neurons in the hippocampus are activated according to the specific location of the animals within the maze (O'Keefe 1976; Ranch 1983). Being hippocampus-dependent does not contradict the fact that other regions in the brain also affect this behavior. For instance, lesions of the cortex (Kolb et al. 1983) or even cerebellum (Goldwitz and Koch 1986) alter the maze performance. However, the central role of the hippocampus and its afferents and efferents in this behavior is well established. Animals with hippocampal lesions performed poorly in the eight-arm maze, with no evidence of functional recovery (Olton 1979). Most pertinent to the present studies is the fact that restoration of deficits in the eight-arm maze, whether due to lesion (Low et al. 1982), aging (Björklund and Gage 1986), or other etiologies, is now routinely achieved in several laboratories. In our laboratory, we destroyed the cholinergic innervations in the hippocampus of adult intact mice by bilateral injection of the cholinergic neurotoxin AF64A (Walsh et al. 1984) into the hippocampus. While control mice needed 2.7 days to reach criterion in the eight-arm maze, only one of ten animals which received injections of 5 nmol AF64A in each hippocampus reached criterion during the six-day test period. After administration of a lower dose of AF64A (2.5 nmol) the treated animals could reach criterion, although it took them longer than control animals (3.4 days). The results reconfirmed the central role of the cholinergic hippocampal innervations in eight-arm maze performance.

In order to assess possible changes in "hippocampal behaviors" after early exposure to barbiturates, prenatally and neonatally treated mice were tested for spontaneous alternations and performance in the eight-arm and Morris water mazes. Results of the first two tests have been presented (Pick and Yanai 1984, 1985). Briefly, neonatally-treated mice had 18% − 36% fewer alternations than controls. The reduction among prenatally treated mice was 31%, but this effect could be elicited only when the more difficult test, delayed spontaneous alternation, was employed. In eight-arm maze performance, prenatally treated mice took twice as long and neonatally treated took four times longer (in days) than control to reach criterion.

The Morris water maze test was applied on prenatally exposed mice as an additional measure for testing early phenobarbital-induced hippocampal behavioral deficits (unpublished observation). The apparatus consisted of a pool (60 cm in diameter; 45 cm deep) filled to a depth of 9 cm with water, made opaque by adding powdered milk (Morris 1984). A transparent glass platform (5 × 5 × 8 cm) was placed into the pool in a constant position, 1 cm under the water surface. In the *place test,* the mice were given two blocks of four trials on each day for four consecutive days, and one block of four trials on the fifth (Nilsson et al. 1987). For each trial the mouse was given 60 s to swim, find the platform and climb on it, then it was allowed to stay and rest for 15 s. Latency, the amount of time in seconds it took the mice to swim from the starting point to the platform, was recorded. On the fifth day, a *spatial probe test* was applied. The platform was removed, the mouse was placed in one of the starting points and its swimming time (in seconds) towards the area of the platform was measured.

In the place test, treated animals were 40% slower than the control animals in the time it took them to reach the platform. In the spatial probe test, the treated animals needed 12% longer than controls to reach the site where the platform was placed before its removal. The results supported the hypothesis on early phenobarbital-induced impairment of hippocampal behaviors with an additional hippocampal behavior, the Morris water maze performance.

Morphological Alterations

These studies were expected to provide background information related to the early phenobarbital-induced biochemical alterations and the corresponding behavioral deficits, since emphasis was placed on areas which are the target innervation areas of the major hippocampal pathways.

The initial studies were designed to examine, in the hippocampus (as well as in other regions), deficits in the area of the different layers and in cell number. The results have been presented previously (Yanai et al. 1979; Yanai and Bergman 1981) and will be mentioned here only briefly. After prenatal exposure to phenobarbital, the adult offspring had fewer prenatally formed neurons than controls, including the hippocampal pyramidal cells. Autoradiographic studies on the treated animals showed a marked reduction in the number of labelled hippocampal pyramidal cells, suggesting that phenobarbital-induced alterations in the time of origin of these neurons (Yanai et al. 1982). The areal parameters appeared unaffected by prenatal phenobarbital administration. Among the neonatally treated offspring the deficit was more severe and also included a deficit in the number of neonatally formed granule cells and a deficit in the area of hippocampal layers.

Further extensive study was conducted on the dendritic tree of the surviving cells (manuscript in preparation). Hippocampal Golgi-stained pyramidal and granule cells were analyzed in prenatally and neonatally exposed mice at postnatal days 14, 21 and 50. Special emphasis was placed on areas presumed to be sites of innervation of the major pathways to the hippocampus. The following parameters were used to assess the metric and the topological characteristics of the neurons: maximal width, length, and area of the dendritic tree; number of dendrites, bifurcation nodes, intersections with concentric circles, and primary dendrites; density of dendrites and spines; and fission angles. The cells of the treated mice showed inferior dendritic arborization to control level, which could be detected mostly at age 50 days. These differences were in the number of intersections of the pyramidal cells with the concentric circles, in their area, fission angles, spine density, number of dendrites and bifurcation nodes. In the granule cells there were differences in the spine density. Fewer changes of the same type i.e., in the number of dendrites and bifurcation nodes in the pyramidal cells and in the branch angles in the granule cells, were observed in the group of neonatal mice at age 21 days. In the 14-day group, the granule cells of the treated mice had significantly lower values than controls for the parameter number of dendrites and bifurcation nodes, while the treated pyramidal cells had significantly higher values than controls in the area of the dendritic tree. The density of dendrites was similar in treated and control cells in all groups. The results suggest long-term deficits in the dendritic architecture of the hippocampal neurons. Many of the deficits corresponded, although not exclusively, to the site of innervation of the septohippocampal cholinergic pathway.

Neurochemical Alterations

These studies were designed to ascertain whether alterations in major hippocampal innervations are the mechanism underlying the effect on behavior of early phenobarbital administration.

The major innervations considered were:
1. The septohippocampal cholinergic pathway. This is probably the most relevant pathway for the hippocampal behaviors, as discussed above.
2. The noradrenergic (NE) pathway which originates from locus coeruleus (Moore and Bloom 1979) and is known to have an effect on hippocampal behaviors (Flicker and Geyer 1982).
3. The serotonergic pathway which originates from the raphe (Graeff et al. 1980). There are some indications for its importance for eight-arm maze behavior (Beatty and Rush 1983).
4. The GABA innervations which are intrinsic in the hippocampus (Storm-Mathisen 1977) and have also been implicated in hippocampus-related behaviors (Grecksch and Matthies 1981).

In studies on the noradrenergic changes after prenatal exposure to pheonbarbital, fluorescing NE cells were counted in the locus coeruleus and the level of NE was assayed in the hippocampus of offspring prenatally treated with phenobarbital. As previously described (Yanai and Pick 1987), the treated offspring had fewer NE cells than controls in the locus coeruleus. This deficit was reflected in a similar reduction of NE levels in the hippocampus. Previous studies have demonstrated short-term alteration in whole brain NE (Middaugh et al. 1981a, b) and long-term changes in hypothalamic NE levels (Yanai et al. 1985) after early exposure to phenobarbital.

Tryptophan hydroxylase was studied in the brainstem of offspring prenatally treated with phenobarbital as a marker enzyme for the serotonergic cells innervating the hippocampus (Yanai et al. 1985). No difference in the activity of tryptophan hydroxylase was found between treated and control animals.

The barbiturate effects are partially mediated by their action on the GABA innervations (Study and Barker 1981), probably via their interaction with the receptor complex of GABA (Garrett and Tabakoff 1986). This receptor complex includes binding sites to GABA/benzodiazepine (BDZ)/picrotoxin−barbiturates/chloride ionophores (Olsen 1982). Chronic administration of barbiturates to adult mice reduces the number of GABA and BDZ receptors in the brain (Mohler et al. 1978; Sonawane et al. 1980). Studies employing [3H]-flunitrazepam suggest that mice which were prenatally exposed to phenobarbital had a decrease at birth in the number (B_{max}) of BDZ receptors in the forebrain and cerebellum, but no changes were observed at age 21 days (Garrett and Tabakoff 1986). In the present experiments (manuscript in preparation), possible alternations in receptor binding of GABA and BDZ were examined. Prenatally- and neonatally-treated mice were studied at age 22 or 50 days. Differences could be found only in the neonatally exposed group at age 22 days; the treated group had a 52% increase over control in GABA receptor B_{max} and 45% in BDZ receptor B_{max}. The relevance of these alterations to the behavioral deficits remains questionable, since the biochemical differences were found only after neonatal exposure, while the behavioral deficits were found in both prenatally and neonatally exposed groups. Furthermore, unlike the behavioral deficits, which were long lasting, deficits in the GABA receptor complex were only transient.

Possible phenobarbital-induced changes in the hippocampal cholinergic innervations are being studied extensively in our laboratory. This is due to the importance of acetylcholine as a principal hippocampal neurotransmitter and to its major role in eight-arm maze behavior, which is now well established (see above). The studies include both prenatally and neonatally exposed groups.

Total cholinesterase (ChE) activity [i.e., acetylcholinesterase (AChE) and pseudocholinesterase (P-ChE)] were assayed in the hippocampus of phenobarbital-exposed offspring during development and adulthood. No differences were found among prenatally exposed offspring either during the embryonic period of exposure to the drug (gestation day 19) or during postnatal life. However, administration of phenobarbital during the neonatal period caused a transient decrease in AChE activity (Kleinberger and Yanai 1985).

ChAT was studied in the phenobarbital-exposed offspring as a marker enzyme for cholinergic activity (manucscript in prepration). The enzyme activity was studied in prenatally and neonatally exposed groups and at ages 22, 30, and 50 days. The results suggest that neither prenatal nor neonatal exposure to phenobarbital could alter ChAT activity. Our present studies are focused on possible early phenobarbital-induced changes in other markers of the presynaptic cholinergic activity. Parallel investigations are being conducted to examine possible postsynaptic changes in the phenobarbital-exposed mice.

Preliminary results suggest that prenatal exposure to phenobarbital alters the number (B_{max}) of the hippocampal muscarinic receptors in 50-day-old offspring. The muscarinic receptors and their second messenger, inositol phosphate, are the subjects of our current investigations.

Manipulating the Phenobarbital-Induced Behavioral Birth Defect

The neuromorphological and biochemical findings should be considered only preliminary at this point. Yet, taken together with previous findings by other investigators, they point to the direction for further research on the reversal of the behavioral deficits. Moreover, the reversal studies may confirm (or contradict) the biochemical studies and thus provide more insight regarding the mechanisms mediating the behavioral effects of prenatal phenobarbital administration.

The prospect for reversing the behavioral deficits by manipulating the serotonergic innervation did not appear promising. Although certain involvement of these innervations in hippocampal behaviors was suggested (Batty and Rush 1983), no effect of early phenobarbital administration on serotonin activity could be demonstrated in our studies (Yanai et al. 1985), and the effect found on other components of serotonin activity by other investigators was only transient (Middaugh et al. 1981a,b).

Similarly, although GABAergic interneurons affect the performance in hippo-campal-related behaviors (Grecksch and Matthies 1981), the phenobarbital-induced changes in GABA were only transient and did not always correspond with sensitive periods for phenobarbital effect on eight-arm maze behavior. Further biochemical studies are necessary to determine the feasibility of reversing the behavioral deficits by GABAergic manipulations.

As it is involved in hippocampal behaviors (Flicker and Geyer 1982) and is clearly altered by early phenobarbital administration (Yanai and Pick 1987), the hippocampal noradrenergic innervation is a likely candidate for further studies on the reversal of the behavioral deficits.

Although our findings on early phenobarbital-induced alterations in the septo-hippocampal cholinergic pathway are only preliminary, an attempt to reverse the phenobarbital-induced deficits by cholinergic manipulations appeared pertinent because of the major role of these innervations in eight-arm maze behavior (as dis-cussed above).

Neuronal Grafting

An attempt was made to at least partially reverse the phenobarbital-induced behavioral deficits by transplantation of cholinergic or noradrenergic neurons. A suspension of embryonic (day 15, where plug day = day 0) dissociated NE or cholinergic cells was transplanted into the hippocampus of prenatally-treated and control mice. The NE cells were taken from the locus coeruleus at the level of the pontine flexure and the rhombic lip. The cholinergic cells were taken from the sep-tal diagonal band by taking the ventral forebrain from the rostral border of the hypothalamus caudal to the olfactory bulb. After behavioral testing, the survival of the transplanted cholinergic neurons was verified with acetylcholine peroxidase antiperoxidase immunocytochemistry (PAP) (Geffard et al. 1985), ChAT fluores-cence immunocytochemistry, and AChE histochemistry (Misulis et a. 1987) 24 h after diisopropyl fluorophosphate (DFP) injection (Mellgren and Sebro 1973). NE transplants were verified with histofluorescence. Stained and fluorescing choliner-gic cells and fluorescing noradrenergic cells could be clearly identified in the respective transplanted hippocampi, but not in the sham-transplanted controls. The transplanted animals and the sham-operated controls were tested in the eight-arm maze (Yanai and Pick 1989). Consistent with previous studies (Pick and Yanai 1985), control offspring needed only 2.1 days to reach criterion in the maze, whereas prenatally treated offspring required 5.2 days. However, transplantation of cholinergic cells into the hippocampus of prenatally treated offspring improved their score to 3.3 days. On the third test day, when the differences were most pro-nounced, 75% of controls but none of prenatally treated reached criterion. How-ever, 55% of the mice prenatally treated with cholinergic transplant reached criterion. Transplantation to normal mice (controls) had no effect on the perfor-mance in eight-arm maze. Transplantation of NE cells did not improve the perfor-mance of prenatally treated mice (Yanai and Pick 1989).

Dopaminergic Denervation

The septal dopaminergic innervations originate from the A_{10} area (Costa et al. 1983) and exert an inhibitory influence on the septohippocampal cholinergic pathways. Their innervation with the cholinergic cells is mostly mediated via GABAergic neurons (Robinson et al. 1979). Previous studies showed that destruction of these dopaminergic innervations enhanced the biochemical activity of the septo-hippocampal pathways. Furthermore, hippocampus-related behaviors, including eight-arm maze performance, were also enhanced by the destruction of the dopaminergic innervation to the septum (Galey et al. 1984). Therefore, it appeared pertinent to attempt to ameliorate the impaired behavior of the phenobarbital-exposed mice by disinhibiting the septohippocampal pathways via the destruction of the septal dopaminergic innervations.

Consequently, adult mice which were exposed to phenobarbital during early development and their respective control groups were given bilateral injections of 6-hydroxydopamine (6-OHDA, 1 µg to each side) into the septum and were later tested for performance in the eight-arm maze (manuscript in preparation). The treatment reduced septal dopamine levels by about half. Hippocampal ChAT levels of animals who were treated with 6-OHDA were more than 1.5 times greater than in their respective control groups.

In the eight-arm maze test, after five days of testing, 86% of control mice reached criterion but only 32% of the mice with prenatal phenobarbital exposure reached criterion. However, 6-OHDA injection into phenobarbital-treated mice increased their score to 75%.

On the first days of testing, phenobarbital-exposed mice who received 6-OHDA injections had scores similar to the phenobarbital-exposed mice which did not receive 6-OHDA, which were five-fold lower than the control group. Only on the third day did they start to show an accelerated rate of improvement and gradually approached control level. Control mice with 6-OHDA injection appeared to have higher scores than intact controls for this parameter in the eight-arm maze. However, these differences were small and did not reach statistical significance.

Both transplantation (Yanai and Pick 1989) and 6-OHDA treatment improved the eight-arm maze behavior of animals exposed in utero to phenobarbital. Yet a basic difference exists between the two treatments. Transplantation repaired the phenobarbital-induced damage where it occurred and thus reversed the behavioral deficits. In fact, transplantation to normal animals never improved their behavior. On the other hand, 6-OHDA improved eight-arm maze performance of all animals, normal or impaired. Thus, the difference between transplantation and 6-OHDA administration is the difference between specific and nonspecific treatments. Another basic difference between the outcome of the two treatments is that transplantation improved the behavior of the treated animals almost from the beginning of the testing period, so that the scores of the transplanted barbiturate animals always paralleled control levels (Yanai and Pick 1989). On the other hand, 6-OHDA improved the behavior substantially only after three testing days. It appears, therefore, that 6-OHDA treatment enhances the ability of the impaired animal to improve with experience, while transplantation at least partially normalized the performance of the impaired mice.

Previous studies have demonstrated the disinhibition by 6-OHDA or other dopaminergic manipulations of the septohippocampal cholinergic pathways on both behavioral and neurochemical levels (Galey et al. 1984). Acetylcholine turnover rate (Robinson et al. 1979) and high affinity choline uptake (Galey et al. 1984) served as indicators for the disinhibition of the neurochemical processes. The present study demonstrates a 6-OHDA treatment-induced rise in ChAT activity as additional evidence for the disinhibition of septohippocampal cholinergic activity following destruction of the A_{10}-septal dopaminergic pathways.

Conclusions

Early exposure to phenobarbital induced deficits in several behaviors in mice (see Reinisch and Sanders 1982; Fishman and Yanai 1983 for review), among them hippocampus-related behaviors. Parallel neuromorphological studies demonstrated early phenobarbital-induced deficits in the hippocampus, including deficits in the sites of innervation of the major pathways to the hippocampus. Biochemical studies revealed alterations in some of the major hippocampal innervations, most pertinent to the behavioral deficits in question being noradrenergic and cholinergic alterations. Neuron transplantation of cholinergic but not noradrenergic cells reversed the deficits and, taken together with what is known in the literature regarding the role of the septohippocampal cholinergic pathway in eight-arm maze behavior, also supported the notion that cholinergic rather than noradrenergic alteration is the relevant mechanism mediating the eight-arm maze deficits induced by prenatal phenobarbital administration. The fact that the same deficits can be ameliorated by disinhibition of the septohippocampal cholinergic pathway via dopaminergic denervation provided further support to this central role of the cholinergic innervations in the behavioral deficits studied.

The studies presented here suggest that, when the mechanisms are fairly well understood, certain behavioral birth defects may be reversed even when many other processes are concomitantly impaired. Furthermore, the techniques employed to reverse the deficits may be used as probes to investigate the mechanism of the deficits.

References

Aadland J, Beatty WW, Maki RH (1985) Spatial memory of children and adults assessed in the radial maze. Dev Psychobiol 18:163–172

Backlund EO, Granberg PO, Hamberger B, Knutsson E, Martensson A, Sedvall G, Seiger A, Olson L (1985) Transplantation of adrenal medullary tissue to striatum in Parkinsonism – first clinical trials. J Neurosurg 62:169–173

Beatty WW, Rush JR (1983) Spatial working memory in rats: Effects of monoaminergic antagonists. Pharmacol Biochem Behav 18:7–12

Björklund A, Gage FH (1986) Transplantation of basal forebrain cholinergic neurons in the aged rat brain. Prog Brain Res 70:499–512

Costa E, Panula P, Thompson HK, Chaney DL (1983) The transynaptic regulation of the septal-hippocampal cholinergic neurons. Life Sci 32:165–179

Fishman RHB, Yanai J (1983) Long-lasting effects of early barbiturates on central nervous system and behavior. Neurosci Biobehav Rev 7:19–28

Flicker C, Geyer MA (1982) Behavior during hippocampal microinfusions. I. Norepinephrine and diversive exploration. Brain Res Rev 4:79–103

Galey D, Durkin T, Sifakis G, Jeffard R (1984) Amélioration de conduites spatiales spontanées et acquises après lésion des afférences dopaminergiques septales chez la souris: relations possibles avec l'activité cholinergique hippocampique. CR Acad Sci Paris 299:681–686

Garrett KM, Tabakoff B (1986) Effect of prenatal phenobarbital on benzodiazepine receptor development. J Neurochem 47:1154–1160

Geffard M, McRae-Degueurce A, Sowan ML (1985) Immunocytochemical detection of acetylcholine in the rodent brain. Science 229:477–479

Goldwitz D, Koch J (1986) Performance of normal and neurological mutant mice on radial arm maze and active avoidance tasks. Behav Neurol Biol 46:216–226

Graeff FG, Auinters S, Gray JA (1980) Median raphe stimulation hippocampus theta rhythm and theta induced behavioral inhibiton. Physiol Behav 25:253–261

Grecksch G, Matthies SH (1981) Differential effects of intrahippocampally or systemically applied picrotoxin on memory consolidation in rats. Pharmaco Biochem Behav 14:613–616

Kleinberger N, Yanai J (1985) Early phenobarbital induced alterations in hippocampal acetylcholinesterase activity and behavior. Dev Brain Res 22:113–123

Kolb B, Sutherland RJ, Whishaw IQ (1983) A comparison of the spatial localization in rats. Behav Neurosci 9:13–27

Low WC, Lewis PR, Bunch ST, Dunnett SB, Thomas R, Iversen SD, Björklund A, Stenevi U (1982) Function recovery following neural transplantation of embryonic septal nuclei in adult rats with septohippocampal lesions. Nature 300:260–262

Mellgren SI, Srebro B (1973) Changes in acetylcholinesterase and distribution of degeneration fibres in the hippocampus region after septal lesions in the rat. Brain Res 52:19–36

Middaugh LD, Thomas TN, Simpson LW, Zemp JW (1981a) Effect of prenatal maternal injections of phenobarbital on brain neurotransmitters and behavior of young C57 mice. Neurobehav Toxicol Teratol 3:271–275

Middaugh LD, Simpson LW, Thomas TN, Zemp JW (1981b) Prenatal maternal phenobarbital increases reactivity and retards habituation of mature offspring to environmental stimuli. Psychopharmacology 74:349–352

Misulis MK, Clinton ME, Dettbern WD, Gupta RC (1987) Differences in central and peripheral neural actions between soman and diisopropyl flurophosphate, organophosphorus inhibitors of acetylcholinesterase. Toxicol Appl Pharmacol 89:391–398

Mohler H, Okada T, Emma SJ (1978) Benzodiazepine and neurotransmitter receptor binding in the rat brain after chronic administration of diazepam or phenobarbital. Brain Res 156:391–396

Moore RY, Bloom FE (1979) Central catecholamine neuron systems: Anatomy and physiology of the norepinephrine and epinephrine systems. Annu Rev Neurosci 2:113–125

Morris R (1984) Development of water maze procedure for studying spatial learning in the rat. J Neurosci Methods 11:47–60

Nilsson OG, Shapiro ML, Gage FH, Olton DS, Björklund A (1987) Spatial learning and memory following fimbria-fornix lesion and grafting of fetal septal neurons to the hippocampus. Exp Brain Res 67:195–215

O'Keefe J (1976) Place units in the hippocampus of the freely moving rat. Exp Neurol 51:78–109

Olsen RW (1982) Drug interactions at the GABA receptor-ionophore complex. Annu Rev Pharmacol Toxicol 22:245–277

Olton DS (1979) The function of septohippocampal connections in spatially organized behavior. Ciba Found Symp 58:327–349

Olton DS, Samuelson RJ (1976) Remembrance of places passed: spatial memory in rats. J Exp Psychol [Animal Behav] 2:97–116

Olton DS, Walker JA, Gage FH (1978) Hippocampus connections and spatial discrimination. Brain Res 139:296–308

Perlow MJ, Freed WJ, Hoffer BJ, Seiger A, Olson L, Wyatt RJ (1979) Brain grafts reduce motor abnormalities produced by destruction of nigrostriatal dopamine system. Science 204:643−647

Pick CG, Yanai J (1984) Long-term reduction in spontaneous alterations after early exposure to phenobarbital. Int J Dev Neurosci 2:223−228

Pick CG, Yanai J (1985) Long-term reduction in eight arm maze performance after early exposure to phenobarbital. Int J Dev Neurosci 3:223−227

Ranch JB (1983) Studies on single neurons in dorsal hippocampal formation and septum in unrestained rats. Exp Neurol 41:461−555

Reinisch JM, Sanders SA (1982) Early barbiturate exposure: The brain, sexually dimorphic behavior and learning. Neurosci Biobehav Rev 6:311−319

Roberts WW, Dember WN, Brodwick M (1962) Alternation and exploration in rats with hippocampal lesions. J Comp Physiol Psychol 55:695−700

Robinson S, Malthe-Sorenssen D, Wood P, Commissiong J (1979) Dopaminergic control of the septal-hippocampal cholinergic pathway. J Pharmacol Exp Ther 208:476−485

Sonawane BR, Yaffe SJ, Shapiro BH (1980) Changes in mouse brain diazepam receptor binding after phenobarbital administration. Life Sci 27:1335−1338

Storm-Mathisen J (1977) Localization of transmitter candidates in the brain: The hippocampal formation as a model. Prog Neurobiol 8:119−181

Study RE, Barker JL (1981) Diazepam and (−) pentobarbital: fluctuational analysis reveals different mechanisms for potentiation of aminobutyric acid responses in cultured central neurons. Proc Natl Acad Sci USA 78:7180−7184

Walsh TJ, Tilson HA, Dehaven DL, Mailman RB, Fisher A, Hanin I (1984) AF64A, a cholinergic neurotoxin, selectively depletes acetylcholine in hippocampus and cortex, and produces long-term passive avoidance and radial-arm maze deficits in the rat. Brain Res 321:91−102

Werner R, Wong D (1987) Correction of genetic diabetes insipidus by adult hypothalamic grafts. Transplantation 43:485−488

Yanai J, Bergman A (1981) Neuronal deficits after neonatal exposure to phenobarbital. Exp Neurol 73:199−208

Yanai J, Pick CG (1987) Studies on noradrenergic alternations in relation to early phenobarbital-induced behavioral changes. Int J Dev Neurosci 5:337−344

Yanai J, Pick CG (1989) Neuron transplantation reverses phenobarbital-induced behavioral birth defects in mice. Int J Dev Neurosci (in press)

Yanai J, Rosselli-Austin L, Tabakoff B (1979) Neuronal deficits in mice following prenatal exposure to phenobarbital. Exp Neurol 64:237−244

Yanai J, Wolf M, Feigenbaum JJ (1982) Autoradiographic study of phenobarbital's effect on development of the central nervous system. Exp Neurol 78:437−449

Yanai J, Sze PY, Iser C, Melamed E (1985) Studies on brain monoamine neurotransmitters in mice after prenatal exposure to barbiturate. Pharmacol Biochem Behav 23:215−219

Differential Neural Plasticity of Diffuse Monoaminergic and Point-to-Point Sensory Afferents as Demonstrated by Responses to Target Deprivation and Fetal Neural Transplants

M. Peschanski, F. Nothias, I. Dusart, B. Onteniente, M. Geffard, and O. Isacson

Summary

Two different types of neural systems have recently been individualized in the mammalian CNS by Sotelo and Alvarado-Mallart (1987) on the basis of anatomical features. Diffuse or global systems (monoaminergic and cholinergic) are characterized by the innervation of widespread areas of the brain originating from a very small number of neurons. Point-to-point or specific systems (specific pathways) are, opposingly, characterized by an almost "one-to-one" connection between neurons.

The present studies have been undertaken to determine whether these two anatomically defined systems exhibit differential regenerative capacities in the adult CNS, using two paradigms, the loss of target-neurons and the introduction of new possible target-neurons. The ventrobasal thalamic complex was chosen because it provided a simple anatomical model in which the two systems are clearly analyzable.

Kainic acid was injected in situ into the rat thalamus in order to obtain a complete neuronal loss in the area. Diffuse afferents (noradrenergic and serotonergic) as well as point-to-point projections (from the dorsal column nuclei and from the somatosensory cortex) were studied one month after lesion in the first series of animals. In a second series, animals were reanesthetized while a suspension of thalamic neurons taken from rat embryos (gestational age 15 days) was implanted in the neuron-depleted area. Afferent fibers were studied at various times between 7 and 90 days after grafting using light and electron microscopy. Diffuse systems were labeled, using immunocytochemistry, with specific antibodies raised against norepinephrine or serotonin; point-to-point afferent fibers were labeled anterogradely using wheat germ agglutinin conjugated to histidine-rich protein (WGA-HRP) as an axonal tracer.

Immunoreactive monoaminergic fibers do not exhibit any alteration in either experimental condition. After neuronal grafting, monoaminergic fibers rapidly grow into the transplants and form a network comparable to normal. Point-to-point afferent fibers are, in contrast, profoundly altered under both conditions. After target neuron depletion, some specific fibers dieback; most fibers, however, exhibit for months a peculiar morphological alteration which has been identified as a regenerative growth cone. When fetal neurons are grafted, point-to-point afferent fibers do not grow into the transplants. They can, however, form synaptic contacts with grafted neurons when these cells develop and invade the area of neuronal loss where regenerative growth cones are present.

In conclusion, this series of experiments demonstrates that diffuse monoaminergic and point-to-point specific afferents do exhibit differential responses to the loss and the reintroduction of possible target neurons. Diffuse systems seem to regenerate altogether much better than specific systems, although the latter do have some regenerative capacities in the adult CNS. These differential regenerative capacities may be related to intrinsic constitutive characteristics opposing the two systems. Alternatively, however, the differential responses may be related to the complete — or incomplete — loss of target neurons, since one monoaminergic neuron will still innervate large unlesioned CNS areas following thalamic lesion. The last hypothesis is that there could be a differential regenerative ability between myelinated (point-to-point) and unmyelinated (monoaminergic) fibers.

Introduction

Until the last decade, the progress of most neurodegenerative diseases was inevitable: there was no way to envisage the replacement of dead neurons in the adult brain. However, expanding research concerning neural transplants has recently renewed both the hope for future therapeutics and interest in the biological events occurring during neurodegeneration.

One major focus of this research is the ability of host central nervous structures to welcome the transplanted neurons. In heterotopic neural grafts, where the goal is to provide target neurons with a specific neurotransmitter, fetal neurons are transplanted into the target zone of the previously eliminated host neurons. The problem, then, lies in the construction of graft-to-host functional connections. In these cases the ability of denervated host tissue to welcome outgrowing axons is under study (see references and discussion in Gage and Björklund 1986). When fetal neurons are transplanted homotopically into an area of neuronal loss to directly replace the missing link of a neural chain, the issue is somewhat different. Then, both host-to-graft and graft-to-host specific connections are essential for the reconstruction of the functional circuitry. In these cases, therefore, one of the major questions is: what happens to host afferent fibers when they are deprived of their target neurons?

The Excitotoxically Lesioned Thalamus

The series of studies reviewed here has been carried out to determine the fate of fibers afferent to the somatosensory relay nucleus of the rat thalamus — the ventrobasal complex (VB) — after an excitotoxic lesion of the area. The rat VB is interesting for such a study because its circuitry is particularly simple (see reference and discussion in Peschanski et al. 1983): it contains a homogeneous population of neurons which all send an axon to the somatosensory cortex. There are neither local circuit neurons nor local axonal collaterals. All the neurons are similarly sensitive to the toxic action of kainic acid, and a total neuronal depletion can be obtained. Afferents can be grouped into three categories:

1. monoaminergic (especially noradrenergic and serotonergic) diffusely organized fibers,
2. point-to-point afferents from spinal and brainstem somesthetic nuclei and from the somatosensory cortex, and
3. intrathalamic GABAergic afferents originating from the thalamic reticular nucleus.

In fact, due to the exquisite sensitivity of thalamic reticular neurons to the kainate injection, there are no GABAergic fibers left in the neuron-depleted VB. We have observed in the course of these studies that responses to the loss of target neurons clearly differ between diffuse monoaminergic systems on the one hand and point-to-point afferents of any source on the other, and results for the two categories are, therefore, presented separately.

Point-to-Point Systems

Two major point-to-point afferent systems in the kainate-lesioned right VB were studied anatomically: the somesthetic projection originating from the dorsal column nuclei and the corticothalamic projections corresponding to the hindlimb representation (partially presented in Peschanski and Besson 1987). Both systems were labeled using wheat-germ agglutinin conjugated to HRP (WGA-HRP: 10% in water or tritiated leucine), injected either bilaterally into the dorsal column nuclei or unilaterally into the SI cortex. Visualization of the transported markers was obtained using tetramethyl/benzidine (Mesulam 1978) or BDHC (benzidine dihydrochloride) (Peschanski et al. 1985a) for HRP at the light and electron microscopic levels, respectively, and classical light microscopy autoradiographic techniques for [^{3}H]-leucine.

At the light microscopic level, labeled elements belonging to both systems were present in the kainate-lesioned VB up to 14 months after the total neuronal depletion. Two major points were made:

1. By comparing results obtained at 10, 20, 30, 60, 150, and >400 days after lesion with those obtained in intact VB, it was clear that, although present, both afferent systems exhibited a progressive decline in density. A decrease in the number of labeled fibers was observed as early as 20 days after lesion, and only rare fibers were labeled in the 14-month-old lesioned VB. This decrease was obvious despite a progressive shrinkage of the VB (around 20% at 30 days, up to 70% at 14 months) which certainly produced an overestimation of the density of fibers in the lesioned site.
2. As early as 20 days after kainate injection, there was a major morphological alteration of labeled terminal fibers which exhibited large (up to 25 μm in diameter), darkly stained swellings. These large structures demonstrated irregular shapes and sizes and bore thin protrusions.

At the electron microscopic level, the neuron-depleted area contained many large varicosities, some of which were labeled with WGA-HRP. Two different kinds of varicosities were distinguished. One type was spherical, contained few organelles − chiefly enlarged mitochondria, and neurofilaments − and was totally ensheathed by glial profiles. On this morphological basis, these structures were considered to be related to dying-back afferent fibers. The other kind of varicosity was totally different: it depicted a bulbous but contorted appearance, contained a wealth of organelles − in particular vesicles and tubules of smooth endoplasmic reticulum, slender mitochondria, and microtubules − and mound-like protrusions, unmyelinated axons (Fig. 1), and neurofilament-filled filopodia issued from the main body. Taking into account reports in the literature, these varicosities were tentatively identified as regenerating growth cones. It was possible, though rarely, to observe transitional forms between morphologically normal terminals and lesioned-type varicosities. One of these transitional forms is shown in Figure 2, in which a small terminal making asymmetrical contact with a remaining piece of a membrane, one called a "moustache" contact (Herndon et al. 1980), is in continuity with a swelling of the preterminal portion of the axon which, quite abnormally, contains a wealth of inclusions and vesicles. It is to be noted that, in the first weeks after lesion, while debris is eliminated, unmyelinated axons and filopodia are rarely seen issuing from growth cones. These thin, long profiles seem to develop subsequently, progressively occupying larger portions of the neuron-depleted area.

The presence of putatively regenerating somesthetic afferents in the neuron-depleted area months after the lesion was made encouraged us to check their function (Peschanski et al. 1985a).

Electrophysiological extracellular recordings using glass micropipettes were made of more than a hundred units whose location in the neuron-depleted area was subsequently histologically controlled. Up to four months after lesion, recorded fibers demonstrated response characteristics to somesthetic stimulation consistent with results obtained in intact animals (see ref. in Guilbaud et al. 1980), including a rough calculation of their conduction velocity.

Diffuse Systems

The fate of monoaminergic fibers afferent to the neuron-depleted VB was anatomically analyzed using immunocytochemical techniques with specific antibodies raised against norepinephrine (NE) or serotonin (5-HT) (Geffard et al. 1984, 1986).

At both the light and electron microscopic levels (Fig. 3) results were unexpectedly normal. There was no morphological alteration of fibers and no obvious change of the fiber density, with the shrinkage of VB even suggesting an increase in density which, once taken into account, was not substantiated.

It was not possible to analyze the function of monoaminergic afferents using electrophysiological recordings because of the lack of a specific response. We turned, therefore, to a biochemical approach to judge the activity of these morphologically unaltered fibers. Results were striking in that, in the neuron-depleted

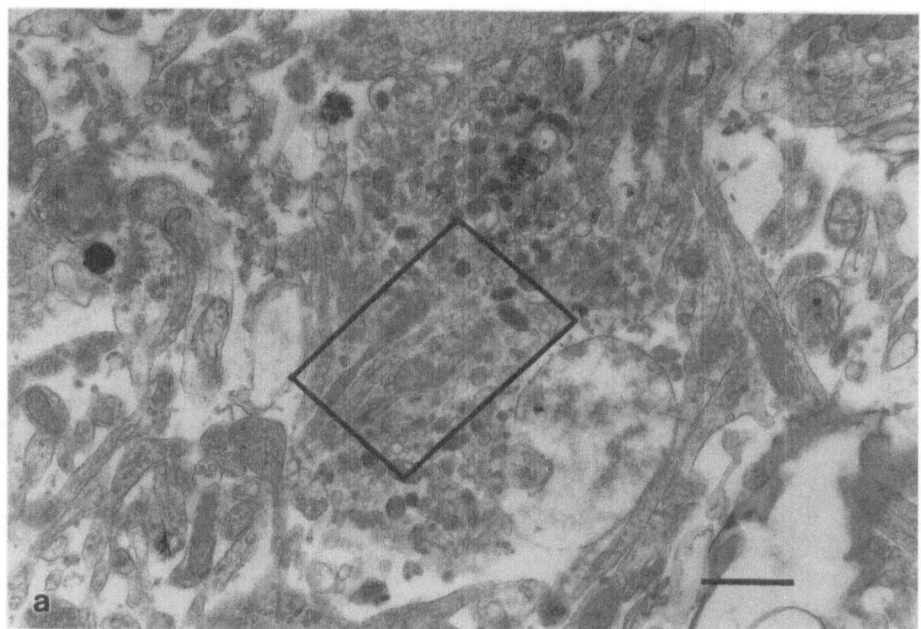

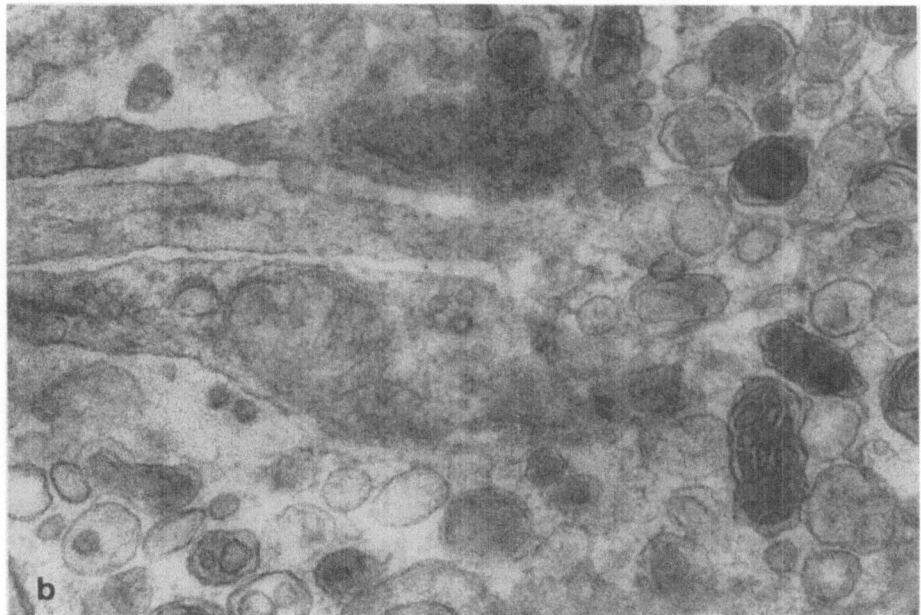

Fig. 1a, b. Electron photomicrographs showing a large regenerating growth cone in the 20-day-old excitotoxically neuron-depleted VB. **b** An enlargement of the area outlined in **a.** *Bar* in **a** is 1 μm, in **b** 0.25 μm

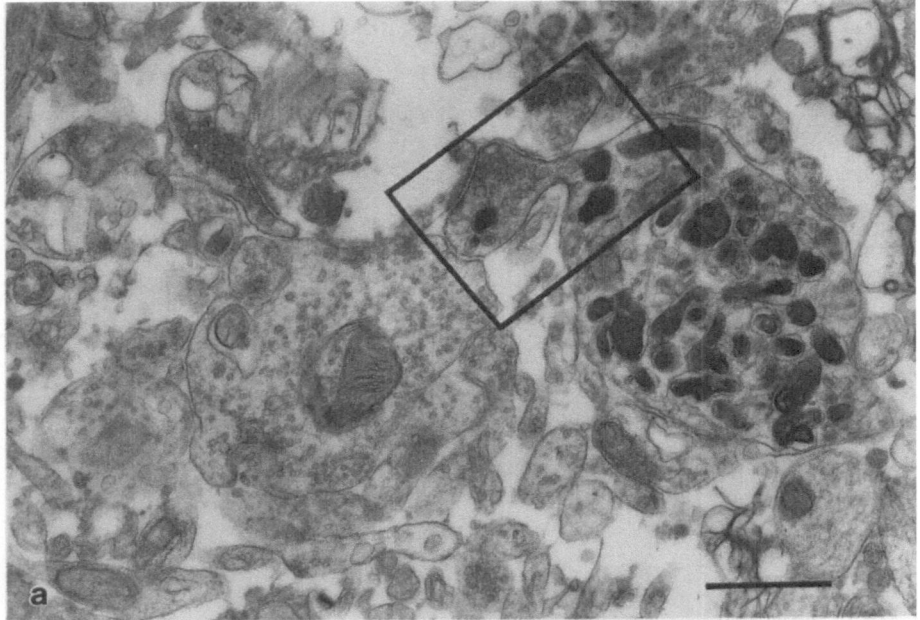

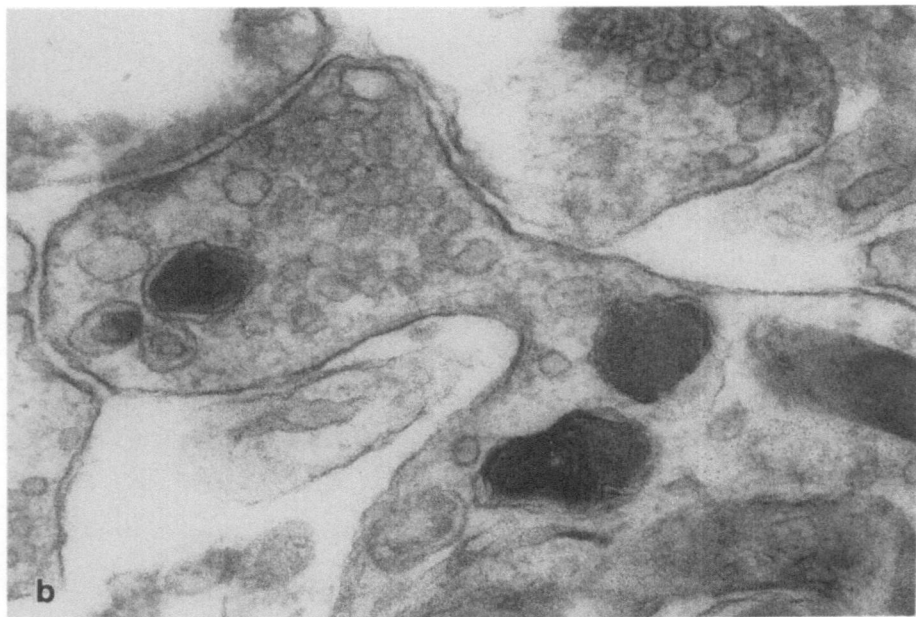

Fig. 2a, b. Electron photomicrographs showing a transitional form observed in the 20-day-old excitotoxically neuron-depleted VB. Note that the small terminal is still attached to a piece of membrane exhibiting remains of a postsynaptic density "moustache". **b** An enlargement of the area outlined in **a**. *Bar* in **a** is 1 μm, in **b** 0.25 μm

VB (up to four months after lesion), no consistent alteration of the total content of any of the monoamines (NE, 5-HT, or dopamine) or of their metabolites could be demonstrated using HPLC and electrochemical detection (Weil-Fugazza et al. 1988).

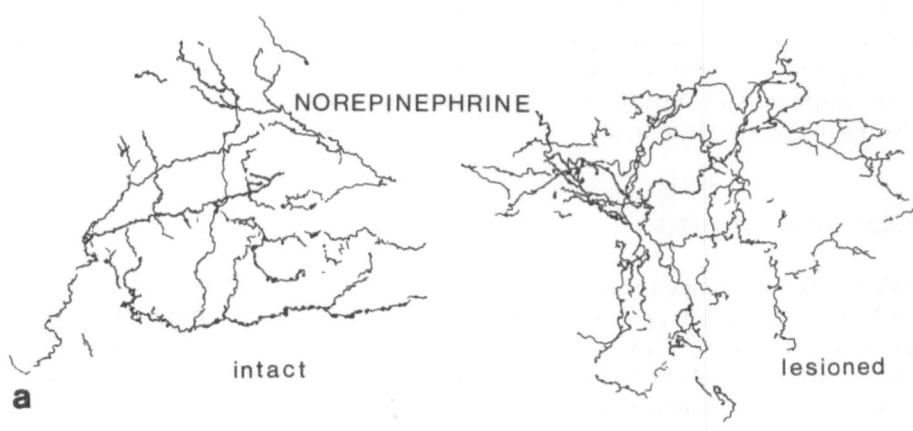

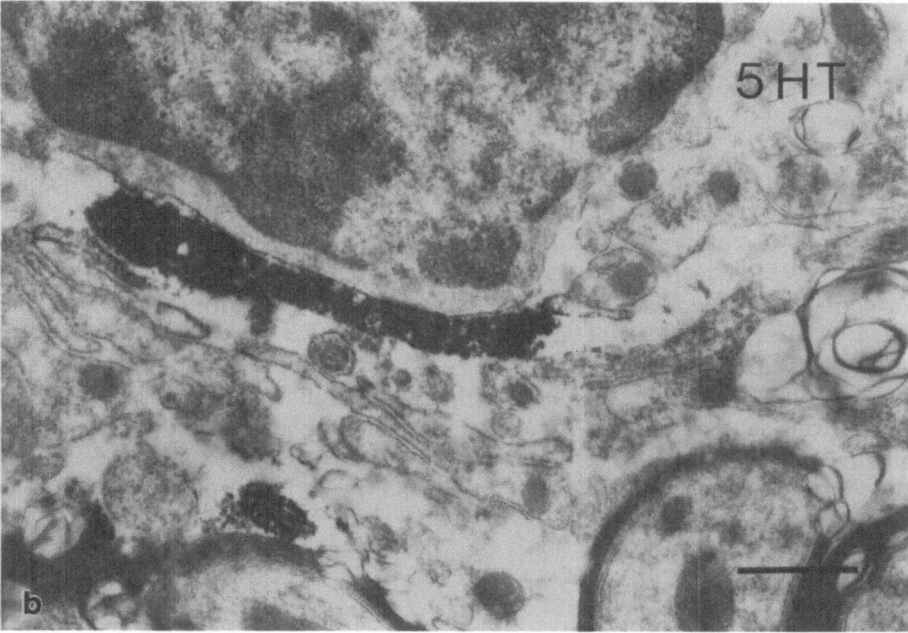

Fig. 3a, b. a Camera lucida drawings showing NE-immunoreactive fibers in the intact and 30-day-old neuron-depleted VB. Note the absence of abnormally large varicosities in the neuron-depleted area. **b** Electron photomicrograph of 5-HT-immunoreactive fibers in the same animal. *Bar* is 1 μm

Responses to Homotopic Fetal Transplants

In the second part of our studies we analyzed the ability of target-deprived afferent fibers to reconstruct a neural circuitry by providing new (transplanted) target neurons. Cell suspension grafts of dissociated thalamic primordia taken from embryos (gestational age 15−16 days) were introduced into the neuron-depleted VB 30 days after kainate injection. This delay was chosen in order to transplant the neurons after the elimination of debris (emptied myelin sheaths are macrophago-cytosed around the beginning of the fourth week after lesion) and when unmyeli-nated processes have started to issue from regenerating growth cones. Again, point-to-point and diffuse systems responded differently to the experimental change in their environment.

Point-to-Point Systems

Afferent fibers labeled from the somatosensory SI cortex or from the dorsal clumn nuclei were present in the transplant two months after transplantation (Peschanski and Isacson 1988). At the light microscopic level, there was a clear difference in the density of dorsal column nuclei projections (WGA-HRP injections were bilat-eral to allow the comparison) between intact and grafted VB. This demonstrated, at the least, the lack of recovery of fiber density and, at the most, the continuation of the progressive fiber loss on the lesion-grafted side. No experimental data can indicate yet whether grafting of new, postsynaptic target neurons prevents fibers from dying back. Labeled fibers were also characterized by their location, avoid-ing the centralmost portion of the transplants, which in our experimental condi-tions essentially coincides with the original transplantation sites.

At the electron microscopic level, point-to-point afferent fibers were morpho-logically different in the area of the transplant where an apparently normal neuropil was reconstructed to those in the marginal zone of the transplant where characteristic features of the neuron-depleted area were still partly visible. In the reconstructed neuropil labeled afferents were rare but occasionally formed synap-tic contacts with dendrites. These synaptic contacts were morphologically similar to those formed by the same afferents in the intact VB ("LR" type terminals for somesthetic afferents, "SR" type terminals for cortical projections). In the margi-nal zone, labeled afferents resembled the regenerating growth cones observed in the neuron-depleted area, although some of them made synaptic contacts with small dendrites. In a recent study, we demonstrated, using [^{14}C]-2-deoxyglucose, that these synaptic contacts allowed the activation of grafted neurons by somesthe-tic stimulation (Juliano et al. 1988).

Diffuse Systems

Noradrenergic and serotonergic fibers originating from the host (most likely from the locus coeruleus and from the midbrain raphe nuclei, respectively) were pre-sent in the transplants as early as six days after injection of the cell suspension.

Later during the growth of the transplant, immunoreactive fibers were present in all portions of the transplants, including the centralmost portion. Up to three months after grafting, however, density of immunoreactive fibers was lower than on the intact side, demonstrating an incomplete constitution of the diffuse afferent networks in the neo-nucleus formed by the transplanted cells. From the morphological point of view, NE and 5-HT fibers did not differ from normal at both the light and electron microscopic levels in 2-week-old and older transplants (thin varicose fibers, Fig. 3). In contrast, 6–12 days after grafting ingrowing monoaminergic fibers exhibited a morphology resembling that of immature monoaminergic fibers (thick, poorly varicose fibers). Experiments designed to analyze the functional state of this innervation have not been carried out yet.

Conclusions and Hypotheses

This series of experiments looking at the fate of afferent fibers when they lose their normal target neurons and when they are provided with new ones is certainly not complete. A number of controls are needed before definite conclusions can be drawn to answer our original question and, further, to hypothesize on what happens to afferent fibers in areas of the human brain depleted in neurons due to a neurodegenerative process. The differentiation between point-to-point and diffuse systems was first elaborated on an anatomical basis by Sotelo and Alvarado-Mallart (1985, 1987), and differential neural plasticity had been hypothesized (1985). Our results strongly support the view of differential neural plasticity of diffuse monoaminergic versus point-to-point sensory afferents in both experimental conditions of responses to target-neuronal loss and to homotypic fetal transplant. There is, for instance, no obvious response of monoaminergic systems to the loss of their VB target neurons. This contrasts with the obvious progressive decline in number of point-to-point afferents and with the formation of putative regenerating growth cones by remaining fibers. When new target neurons are provided by transplantation, responses also differ, as schematized in Figure 4. Monoaminergic afferents grow into the transplant as early as a few days after grafting. This precocity is all the more striking since we have observed (Nothias et al., in preparation) that transplanted fetal neurons form a dense neuropil and synaptogenesis really starts only 8 to 10 days after grafting, i.e., well after monoaminergic afferents have started to innervate the transplant. Innervation of transplanted neurons by point-to-point fibers seems to depend upon another mechanism. Although results are preliminary and further analysis is required, point-to-point fibers do not seem to actually grow into the VB transplant. During the first weeks after grafting, the size of the transplant increases because fetal neurons grow and their dendritic arbors develop. This is demonstrated, in particular, by a parallel increase in transplant size and decrease of cell density over the first month. This increase in size seems to differentiate two parts in the transplant: the original zone of transplantation and the zone of expansion. In the original zone of transplantation a mechanical disruption of the host tissue has been produced during the transplantation. To innervate this area an actual ingrowth of the afferents is needed, and this seems possible only for monoaminergic diffuse systems in VB. In contrast, the zone of transplant

8 days   15 days   30 days

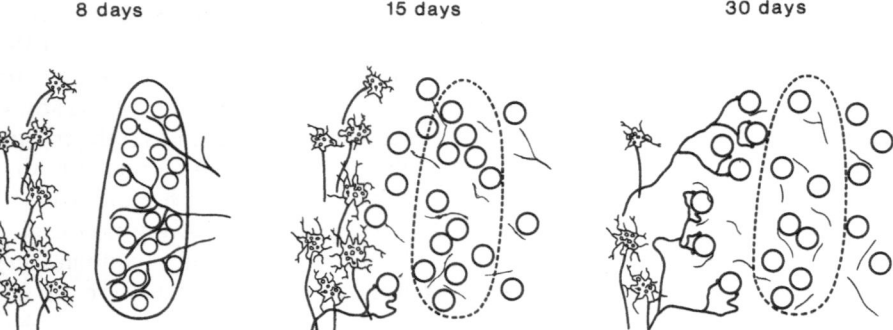

Fig. 4. Schematic drawings representing our hypotheses concerning the formation of connections between transplanted neurons and the two types of afferent systems. The original transplantation zone has been indicated in all three drawings for clarification of the explanation although no obvious cytoarchitectural features permit its precise delineation in most transplants. For the sake of simplicity only, diffuse monoaminergic afferents come from the *right* side of the transplants and point-to-point system from the *left*

expansion is created by the spread of transplanted neurons into the neuron-depleted host tissue. It is then possible for all afferent fibers present in the area – including point-to-point systems – to be in close proximity to fetal neurons and, eventually, to contact them. Such an hypothesis, if it was confirmed, would underline the differential plasticity of the two systems: one capable of actual ingrowth and the other capable only of reconnection with locally available target neurons. It is interesting to note that similar results concerning the differential fate of diffuse monoaminergic and point-to-point afferent fibers have recently also been obtained for the striatum under both of the experimental conditions of neuronal depletion (Nothias et al. 1988a; Walker and McAllister 1986) and homotopic grafts (Wictorin et al. 1988).

Another important result of this series of experiments is the demonstration of a progressive loss – though over a very long time period – of point-to-point afferent fibers which are deprived of their target neurons. We show that morphologically identified degenerating and growing processes coexist in areas of neuron depletion leading to a progressive decrease in the number of afferent fibers. Regenerating growth cones are able (at least some of them) to construct functional connections with new postsynaptic target neurons provided, but degenerated fibers obviously cannot. It is, therefore, conceivable that a "critical period" exists after neuronal loss during which putatively regenerating afferent fibers are still plentiful and the possibility of good integration of transplanted neurons into the host neural circuitry is still offered. Experimental excitotoxic lesions are supposed to mimic some mechanisms of neurodegenerative diseases in humans, although the imperfections of the model are obvious. One of the major differences between neurodegenerative diseases and their experimental models is the duration of the evolution. In neurodegenerative diseases the neuronal loss progresses over years, whereas in experimental conditions neurons are eliminated all at once. The existence of a "critical period" during which transplantation would be more efficient, if it is also observed in neurodegenerative diseases, would then be a major factor to be taken into account for successful therapeutics.

Acknowledgments. The authors wish to thank Dr S. L. Juliano for her help with the manuscript, and F. Roudier, E. Dehausse, L. Legoic, G. Geraud, J. Bureau, and M. Rabat for skillful technical assistance.

References

Gage FH, Björklund A (1986) Enhanced graft survival in the hippocampus following selective denervation. Neuroscience 17:89–98

Geffard M, Seguela P, Buijs RM (1984) Immunorecognition of antiserotonin antibodies by using a radiolabelled ligand. Neurosci Lett 50:217-222

Geffard M, Patel S, Dulluc J, Rock AM (1986) Specific detection of noradrenaline in the rat brain using antibodies. Brain Res 363:395–400

Guilbaud G, Peschanski M, Gautron M, Binder D (1980) Neurones responding to noxious stimulation in VB complex and caudal adjacent regions in the thalamus of the rat. Pain 8:303–318

Herndon RM, Addick E, Coyle JT (1980) Ultrastructural analysis of kainic acid lesion to the cerebellar cortex. Neuroscience 5:1015–1026

Juliano SL, Dusart I, Peschanski M (1988) A metabolic and histochemical evaluation of fetal homotypic transplants into excitotoxically lesioned rat ventrobasal complex. In: Bentivoglio M, Spreafico R (eds) Cellular mechanisms in the thalamus. Elsevier, Amsterdam

Mesulam MM (1978) Tetramethyl benzidine for horseradish peroxidase neurohistochemistry: a non carcinogenic blue reaction product with superior sensitivity for visualizing neural afferents and efferents. J Histochem Cytochem 26:106–117

Nothias F, Onteniente B, Geffard M, Peschanski M (1988a) Rapid growth of host afferents into fetal thalamic transplants. Brain Res 463:341–345

Nothias F, Wictorin K, Isacson O, Björklund A, Peschanski M (1988b) Morphological alteration of thalamic afferents in the excitotoxically lesioned striatum. Brain Res 461:349–354

Peschanski M, Besson JM (1987) Structural alteration and possible growth of afferents after kainate lesion in the adult rat thalamus. J Comp Neurol 258:185–203

Peschanski M, Isacson O (1988) Fetal homotypic transplant in the excitotoxically neuron-depleted thalamus: light microscopy. J Comp Neurol 274:449–463

Peschanski M, Guilbaud G, Lee CL, Mantyh PW (1983) Involvement of the ventrobasal complex of the rat thalamus in the sensory-discriminative aspects of pain: electrophysiological and anatomical data. In: Macchi G, Rustioni A, Spreafico R (eds) Somatosensory integration in the thalamus. Elsevier, Amsterdam, pp 147–163

Peschanski M, Roudier F, Ralston HJ III, Besson JM (1985a) Ultrastructural analysis of the terminals of various somatosensory pathways in the ventrobasal complex of the rat thalamus: an electron-microscopic study using wheat-germ agglutinin conjugated to horseradish peroxidase as an axonal tracer. Somatosens Res 3:75–87

Peschanski M, Briand A, Poingt JP, Guilbaud G (1985b) Electrophysiological properties of lemniscal afferents in rat after kainic acid lesions in the ventrobasal thalamus. Neurosci Lett 58:287–292

Sotelo C, Alvarado-Mallart RM (1985) Cerebellar transplants: immunocytochemical study of the specificity of Purkinje cell inputs and outputs. In: Björklund A, Stenevi U (eds) Neural grafting in the mammalian CNS. Elsevier, Amsterdam, pp 205–215

Sotelo C, Alvarado-Mallart RM (1987) Reconstruction of the defective cerebellar circuitry in adult pcd mutant mice by Purkinje cell replacement through transplantation of solid embryonic implants. Neuroscience 20:1–22

Walter PD, McAllister JP (1986) Anterograde transport of horseradish peroxidase in the nigro-striatal pathway after neostriatal kainic acid lesion. Exp Neurol 93:334–347

Weil-Fugazza J, Peschanski M, Godefroy F, Manceau V, Besson JM (1988) Absence of long-term changes in biochemical markers of monoaminergic systems afferent to the excitotoxically neuron-depleted somatosensory thalamus. Brain Res 444:374–379

Wictorin K, Isacson O, Fischer W, Nothias F, Peschanski M, Björklund A (1988) Connectivity of striatal grafts implanted into the ibotenic acid-lesioned striatum. I. Subcortical afferents. Neuroscience 27:547–562

Deleterious and "Overshoot" Effects of Intracerebral Transplants

B. Will, J.-Ch. Cassel, and *C. Kelche*

Summary

Behavioral and other abnormalities, as well as neurochemical "overshoot" effects, have been found in studies on intracerebral transplants. "Overshoot" effects have been observed in brains with grafts involving cholinergic, serotonergic, or noradrenergic systems. Although few in number, these reports raise the question of the risks of graft-induced neurotoxicity which may explain some of the graft-induced impairments. This question should be carefully addressed before accepting the technique of intracerebral grafting for clinical application.

"Overshoot" Effects of Intracerebral Transplants

Over the last decade, substantial evidence has accumulated showing that intracerebral grafts of fetal tissue are able to promote structural and functional recovery in animals with brain damage of either surgical, genetic or other origin (Azmitia and Björklund 1987; Björklund and Stenevi 1985). Although many successful studies have led neurobiologists to a rather optimistic view of intracerebral grafting techniques, "brain grafters" are sometimes plagued by results showing that grafts may fail to induce the expected recovery from brain damage or even that they may induce behavioral and other abnormalities.

In some cases, graft-induced behavioral impairments were actually larger than those induced by lesions alone. For instance, abnormally high spontaneous locomotor activity and abnormally low contact time with novel objects in an open field test were reported by Dunnett et al. (1982) in rats with fimbria-fornix lesions and septal grafts (Fig. 1). Graft-induced impairments additive to lesion-induced impairments have also been found in cognitive performance of rats with intrahippocampal grafts of fetal septal cell suspensions and selective fimbria-fornix lesions (Dalrymple-Alford et al. 1987) (Fig. 2) and in rats with intrahippocampal grafts of fetal hippocampal tissue and hippocampal lesions (Woodruff et al. 1988). Moreover, the susceptibility to pentylenetetrazol-induced seizures was found to be enhanced (although the intensity of audiogenic seizures was decreased) in rats with intrahippocampal septal transplants and selective fimbria-fornix lesions (Cassel et al. 1987) (Fig. 3). In line with this finding, Buzsáki and Gage (this volume)

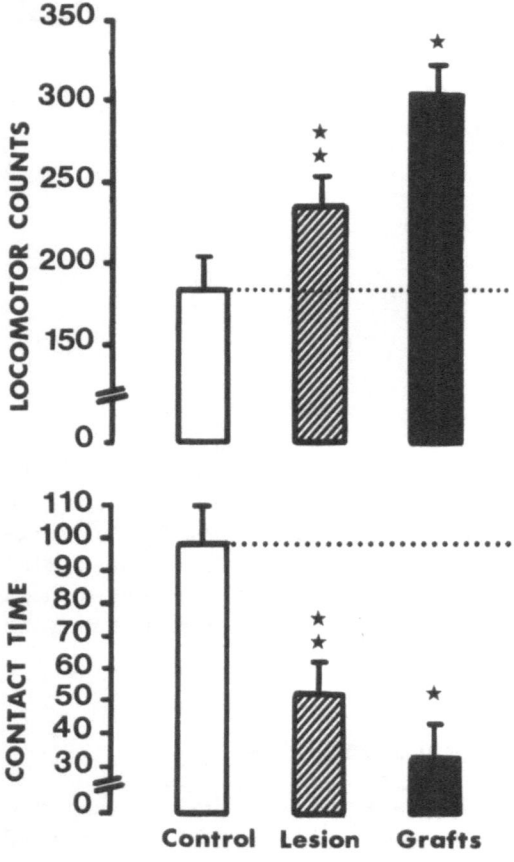

Fig. 1. Mean number of the daily locomotor counts (5 days) and of the daily contact time with a novel object (on days 4 and 5) in an open field. Subjects were rats with fimbria-fornix aspiration lesions and fetal septal cell transplants placed either in the lesion cavity as tissue blocs or in the hippocampus as a cell suspension. Performances of nongrafted rats with lesions differed significantly from those of contral rats ($p < .01$). Performances of grafted rats differed significantly from those of rats with lesion ($p < .05$) and those of control rats ($p < .01$). (Results from Dunnett et al. 1982)

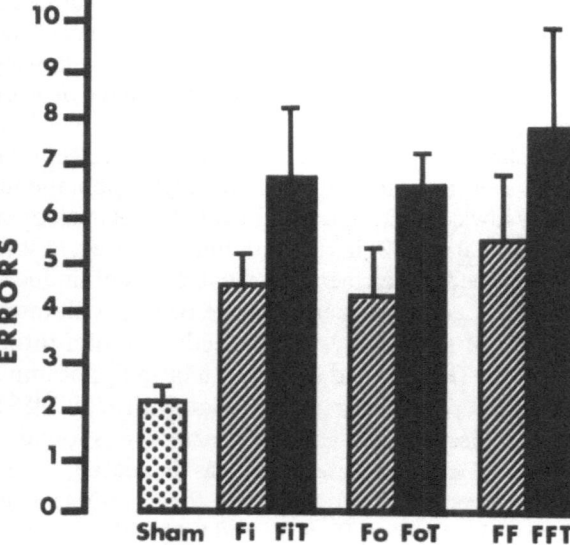

Fig. 2. Mean number of errors recorded 7 months after transplant surgery in an eight-arm radial maze. Transplantation consisted of injecting a fetal cell suspension into the hippocampus (2 sites per hippocampus; 2 µl/site) in rats with partial destruction of the fimbria-fornix pathways. *Sham,* sham-operation; *Fi,* medial fimbria lesion; *FiT,* Fi lesion + transplant; *Fo,* dorsal fornix lesion; *FoT,* Fo lesion + transplant; *FF,* Fi + Fo lesion; *FFT,* Fi + Fo lesion + transplant. There was a significant overall deleterious effect of the graft ($p < .001$)

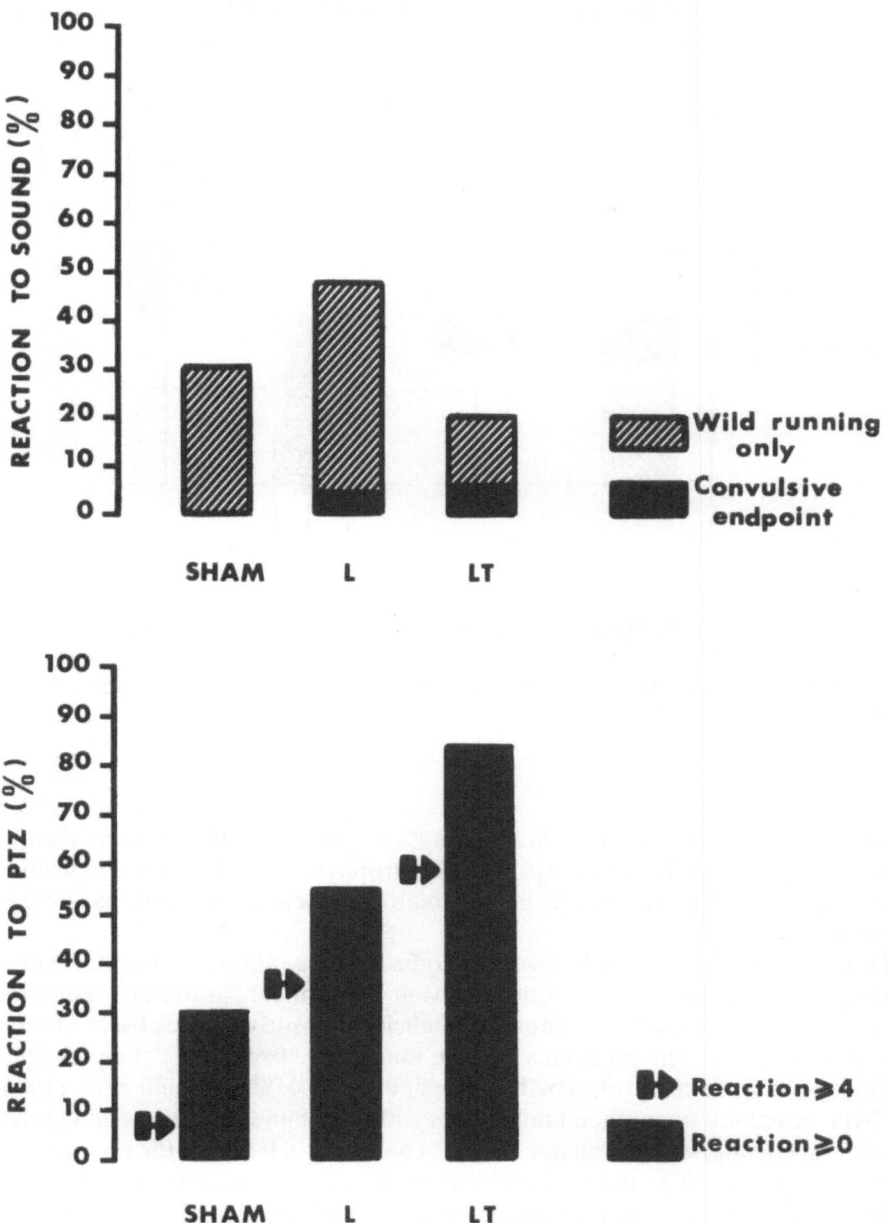

Fig. 3. *Upper part,* Audiogenic susceptibility, or reactivity to sound (10- to 20-kHz peaks, 120 dB for 90 s), given as the percentage of rats responding with at least one wild running phase (*hatched* areas) or reaching convulsive endpoint (*black* areas). *Lower part* Reactivity to pentylenetetrazol (PTZ) (30 mg/kg, i.p.). *Reaction* ≥ *4* indicates a generalized reaction (head, forelimb and hindlimb jerks at least, paroxysmal convulsive attack followed by catatonia); *Reaction* ≥ *0* includes also the less severe reactions. *SHAM,* sham-operated; *L,* rats with partial lesions of the fimbria-fornix (Fi, Fo, FF; see Fig. 2); *LT,* rats with partial fimbria-fornix lesions and intrahippocampal cell suspension grafts

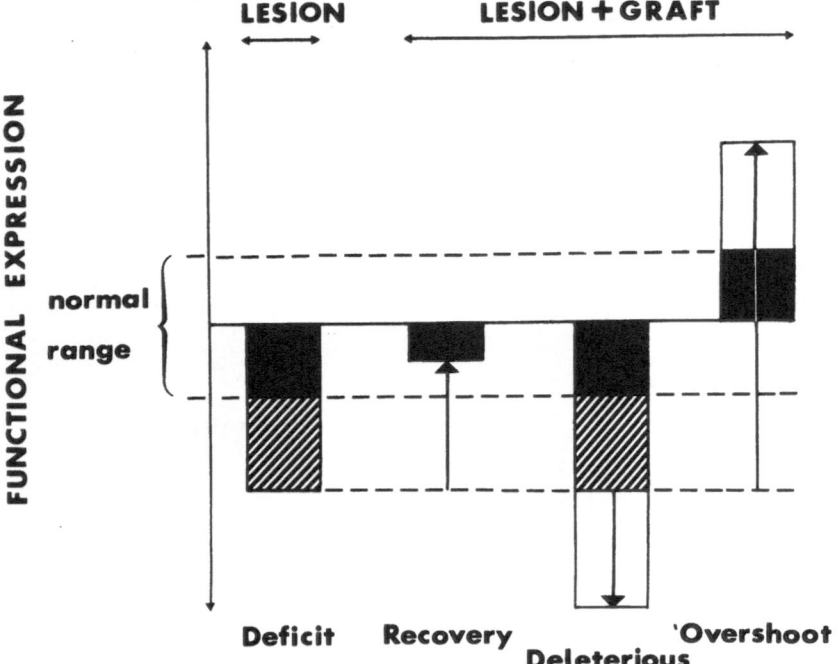

Fig. 4. Schematic representation of various possible effects of intracerebral grafts on lesion-induced abnormalities

observed grand mal symptoms in about 50% of the rats with intrahippocampal bilateral hippocampal grafts and provided eletrophysiological evidence indicating that hippocampal grafts should be regarded as potential epileptic pacemaker sources.

However, there is yet another way for grafts to induce abnormalities: Instead of adding an impairment to the lesion-induced impairment, grafts may generate excessive recovery, leading eventually to deficits opposite to those observed without grafts (Fig. 4). The possibility of such functional "overshoots" may be illustrated by recent data reported by Herman et al. (1985); and Choulli et al. (1986) on hyperreactivity to amphetamine in rats with dopaminergic lesions and grafts. Ipsilateral circling, which is characteristic of a unilateral lesion of the nigrostriatal pathway, gave way in grafted animals to contralateral circling, and amphetamine-induced locomotor activity was restored in animals with bilateral lesions and grafts to levels above those observed in control animals (Fig. 5).

Graft-induced abnormalities may be explained by several factors, such as obstruction of the free flow of cerebrospinal fluid resulting in hydrocephalus, mechanical compression of host neural tissue, alteration of the blood-brain barrier, abnormal graft-to-host connections, or intragraft reverberatory excitatory circuits (Buzsáki and Gage 1988). One additional possibility may be considered, namely graft-induced overproduction of neurotransmitters. This possibility,

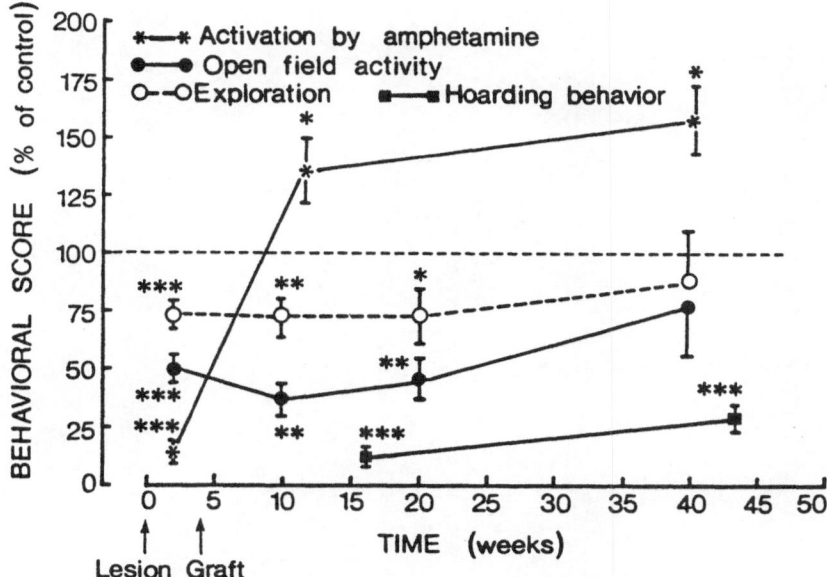

Fig. 5. Time course of behavioral recovery (*filled circles*, open field activity; *open circles*, exploration; *filled squares*, hoarding behavior; *stars*, amphetamine-induced activation) given as a percentage of control performances, in rats with 6-OHDA-induced lesions of the nucleus accumbens and intraaccumbens grafts of fetal mesencephalic cells. Results are mean ± SEM; significantly different from control: * $p < .05$; ** $p < .01$; *** $p < .001$. (From Herman et al. 1985)

although mentioned briefly in a few reports, has not yet been sufficiently emphasized.

Grafts may induce levels of recovery of certain indices of neurotransmission which far exceed control levels observed in intact animals. For instance, in rats with aspiration lesions of the dorsal septohippocampal pathways, Björklund et al. (1983) showed that intrahippocampal fetal septal suspension implants promoted recovery of hippocampal choline acetyltransferase (ChAT) activity within 6 months after lesion surgery. Graft-derived ChAT activity increased within 3 weeks in the vicinity of the grafts and expanded over the entire hippocampus within 6 months. As shown in their Figure 2, 6 months after grafting the mean ChAT activity levels reached about 100%–120% of normal in the hippocampal slices located near the grafts; however, some individual values reached about 170% of control in these slices, a value far beyond the normal range. In another recent study carried out in rats with partially deafferented hippocampi and intra-hippocampal fetal septal suspension grafts. Dalrymple-Alford et al. (1987) found that hippocampal graft-derived acetylcholinesterase (AChE) positivity was more widely distributed in some rats than that observed in rats with intact brains (Fig. 6). In the area CA1 the densest AChE positivity, which is normally located close to pyramidal cell bodies, had clearly expanded over all layers. Further, graft-

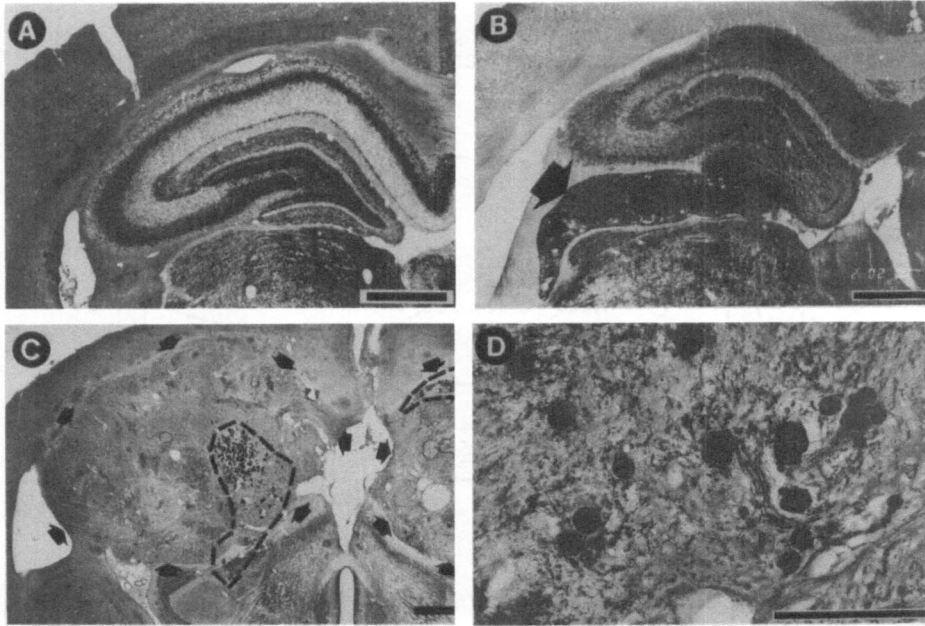

Fig. 6A–D. Coronal sections through the AChE-stained dorsal hippocampus of a sham-operated rat **(A)** and of rats which received intrahippocampal fetal septal cell transplants 10 days following electrolytic lesions of the medial fimbria **(B–D).** Note the excessive AChE density in the dorsal hippocampus shown in **B** as compared to **A,** the hypertrophied graft and the resulting destruction of the dorsal hippocampus in **C,** and the hyaline structures in **C** *(interrupted lines)* and **D.** *Arrows* indicate the location of the grafts. *Scale bars:* 1 mm in **A–C;** 250 μm in **D**

derived AChE positivity was also found to be denser in the dentate gyrus than that found in rats with intact brains.

A similar "neurochemical overcompensation" has been observed in the serotonergic system as well. Auerbach et al. (1985) showed that intrahippocampal minced fetal raphe nuclei implants given 15 days after neurotoxic 5,7-dihydroxy-tryptamine (DHT)-induced lesions of the rostral raphe nuclei in adult rats "restored" hippocampal serotonin (5-HT) and 5-hydroxyindolacetic acid (5-HIAA) concentrations to *average levels,* which in fact largely exceeded those observed in intact rats (about 125% – 150% of normal 5-HT and 5-HIAA levels). Zhou et al. (1987) found similar results in rats which received 5,7-DHT-induced lesions of the fimbria-fornix pathways and, 2 weeks later, minced or dissociated fetal cells of either the raphe or locus coeruleus nuclei. One month following graft-ing of raphe cells, 5-HT levels as well as synaptosomal high affinity uptake of (3H)-5-HT were restored to average levels which largely exceeded those observed in intact rats (Fig. 7). In rats which received grafts of locus coeruleus cells, noradren-aline levels were restored above control levels only when these cells were trans-planted into intact recipients but not when they were transplanted into the hippo-

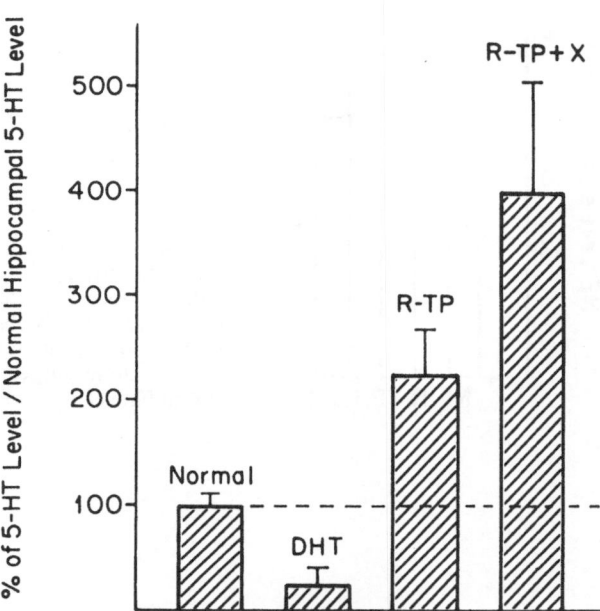

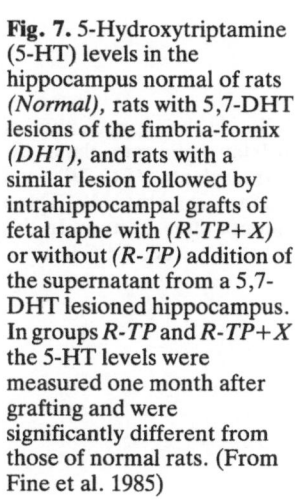

Fig. 7. 5-Hydroxytriptamine (5-HT) levels in the hippocampus normal of rats *(Normal)*, rats with 5,7-DHT lesions of the fimbria-fornix *(DHT)*, and rats with a similar lesion followed by intrahippocampal grafts of fetal raphe with *(R-TP+X)* or without *(R-TP)* addition of the supernatant from a 5,7-DHT lesioned hippocampus. In groups *R-TP* and *R-TP+X* the 5-HT levels were measured one month after grafting and were significantly different from those of normal rats. (From Fine et al. 1985)

campus of rats with 5,7-DHT-induced lesions of the fimbria-fornix pathways (Fig. 8). However, when neural tissue is grafted into an intact host brain, transplanted neurons may co-exist with intrinsic terminals. Therefore, the local "overshoot" observed may actually be due to the addition of both host- and graft-derived products, even if those derived from graft remain below normal levels. In another recent study, Zetterström et al. (1986) found that in rats which received 6-hydroxydopamine (6-OHDA) lesions of the mesostriatal dopaminergic pathway followed by intrastriatal grafts of fetal mesencephalic dopaminergic neurons, the levels of 5-HIAA measured in the vicinity of the grafts by brain dialysis were increased in comparison to those found in the striata of nongrafted animals. Strecker et al. (1987) also found that 5-HIAA levels were higher in rats with unilateral lesions of the nigrostriatal dopamine pathways and intrastriatal grafts of dopamine-rich cell suspensions than in intact animals (Fig. 9).

These few reports of a graft-induced "overshoot" effect in the recovery of certain indices of neurotransmission concern the cholinergic, serotonergic and noradrenergic systems. Regardless of whether this phenomenon is restricted to these neuronal systems, such excessive effects might be indicative of graft-derived neuronal hyperactivity in the host or in the graft itself which, in turn, could affect other neuronal systems. One might hypothesize that such a graft-induced excess of neurotransmitters may result in a cascade of neurochemical events including downregulations of both pre- and postsynaptic receptor activity as well as changes in turnover and enzyme induction in the host brain. Such reactions can be considered as adaptive since they may counterbalance, at least partially, the effects of a chronic excess of substances involved in neurotransmission. However, graftinduced downregulations may be insufficient to prevent excitotoxicity which may

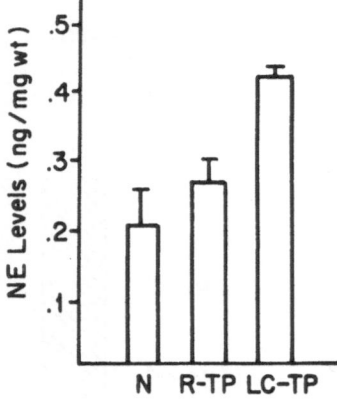

Fig. 8. Levels of noradrenaline in the hippocampus of rats with intrahippocampal fetal raphe transplants *(R-TP)* or locus coeruleus transplants *(LC-TP)* and compared to levels observed in normal rats *(N)*. These levels were measured one month after grafting. In this case, grafting surgery was not preceded by lesions. (From Fine et al. 1985)

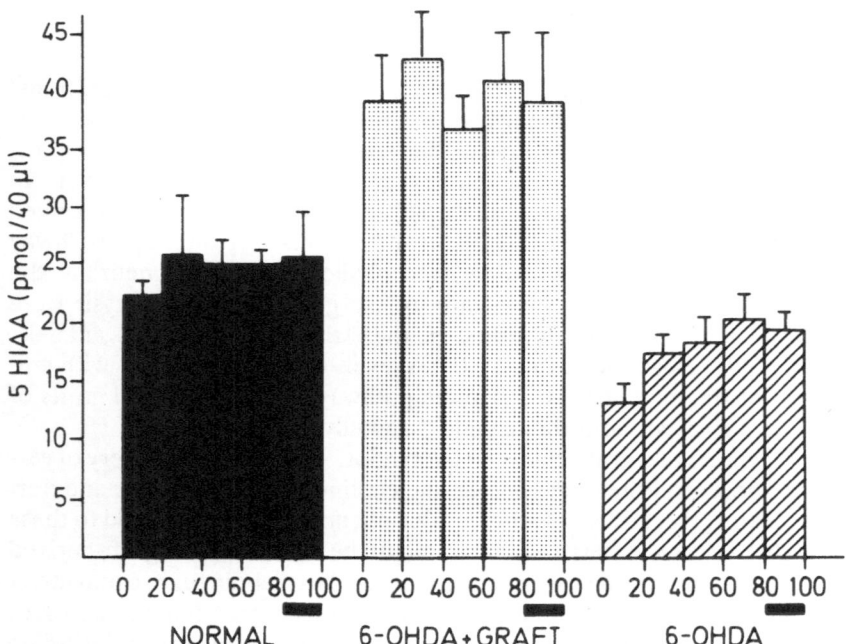

Fig. 9. Levels of 5-HIAA recovered in the dialysis perfusates from normal intact striata (*black* bars), 6-OHDA-lesioned and grafted striata (*strippled* bars), and 6-OHDA-lesioned striata without grafts (*hatched* bars). The four 20-min samples show basal measurements. For the fifth sample (80–100, *soli horizontal bar*), 10^{-5} M D-amphetamine was included in the perfusion medium. Levels of 5-HIAA of rats with lesions + grafts were significantly increased as compared to those of normal rats or rats with lesions only. (From Turski et al. 1983)

result directly from locally excessive amounts of excitatory amino acids or indirectly from locally excessive amounts of excitatory amino acids or indirectly from locally excessive amounts of transmitters that produce *in fine*, via one or several interneurons, an excessive release of excitatory amino acids from a target neuron. It is known that excitotoxic damage to limbic structures such as the hippocampus or the amygdala can be induced by intracerebral microinjections of cholinomimetics at doses which cause supraphysiological stimulation of limbic muscarinic receptors (Turski et al. 1983). Consequently, it can be further hypothesized that "overactive" grafts, especially intrahippocampal septal grafts which are assumed to release acetylcholine into the host structure, may be compared with generators of supraphysiological stimulation, which may in turn induce excitotoxic damage to the host brain. In relation to this hypothesis, we have recently obtained (Dalrymple-Alford et al. 1987; Cassel et al. 1987) histological data which suggest the possible occurrence of intrahippocampal graft-derived excitotoxicity:

1. In some rats, grafts which had completely destroyed the dorsal hippocampus contained dense accretions of spherical, hyaline structures (Fig. 6) similar to those described by Fine et al. (1985) as corresponding most probably to the final stage of tissue degeneration induced by excitotoxic processes.
2. In some other rats, extensive parts of the hippocampus were destroyed in the vicinity of very small grafts. This destruction was consequent neither to graft-induced compression nor to the partial fimbria-fornix lesions that were made ten days prior to grafting.

Although addressed to date by a limited number of studies, excitotoxic damage to brain tissue subsequent to graft-induced excess of neurotransmitters in some brain areas remains a potentially important side-effect of intracerebral transplants worth considering from both a theoretical and clinical point of view.

Acknowledgments. Our research was supported by grants to BE Will from the Fondation pour la Recherche Médicale and from the Institut National de la Santé et de la Recherche Médicale (866.019). We thank Drs. C. Snead, S. Brailowsky, and J. Anderson for reviewing the manuscript.

References

Auerbach S, Zhou F, Jacobs BL, Azmitia E (1985) Serotonin turnover in raphe neurons transplanted into rat hippocampus. Neurosci Lett 61:147–152

Azmitia E, Björklund A (eds) (1987) Cell and tissue transplantation into the adult brain. Ann NY Acad Sci 495:1–813

Björklund A, Stenevi U (1985) Neural grafting in the mammalian CNS. Elsevier, Amsterdam, pp 1–709

Björklund A, Gage FH, Schmidt RH, Stenevi U, Dunnett SB (1983) Intracerebral grafting of neuronal cell suspensions. VII. Recovery of choline acetyltransferase activity and acetylcholinesterase synthesis in the denervated hippocampus reinnervated by septal suspension implants. Acta Physiol Scand [Suppl] 522:59–66

Buzsáki G, Gage FH (1988) Neural grafts: possible mechanisms of action. In: Petit TL (ed) Neural plasticity: a lifespan approach. Liss, New York

Cassel JC, Kelche C, Will BE (1987) Susceptibility to pentylenetetrazol-induced and audiogenic seizures in rats with selective fimbria-fornix lesions and intrahippocampal septal grafts. Exp Neurol 97:564–576

Choulli K, Herman JP, Rivet JM, Simon H, Le Moal M (1986) Behavioral recovery following 6-OHDA lesions of the nucleus accumbens and intra-accumbens implantations of dopaminergic grafts. In: Briley H, Kato A, Weber M (eds) New concepts in Alzheimer's disease. Macmillan, New York, pp 265–279

Dalrymple-Alford JC, Kelche C, Cassel JC, Toniolo G, Pallage V, Will BE (1987) Behavioral deficits after intrahippocampal fetal septal grafts in rats with selective fimbria-fornix lesions. Exp Brain Res 69:1–15

Dunnett SB, Low WC, Iversen SD, Stenevi U, Björklund A (1982) Septal transplants restore maze learning in rats with fimbria-fornix lesions. Brain Res 251:335–334

Fine A, Dunnett SB, Björklund A, Clarke D, Iversen SD (1985) Transplantation of embryonic ventral forebrain neurons to the neocortex of rats with lesions of nucleus basalis magnocellularis. I. Biochemical and anatomical observations. Neuroscience 16:769–786

Herman JP, Choulli K, Le Moal M (1985) Hyper-reactivity to amphetamine in rats with dopaminergic grafts. Exp Brain Res 60:521–526

Strecker RE, Sharp T, Brundin P, Zetterström T, Ungerstedt U, Björklund A (1987) Autoregulation of dopamine release and metabolism by intrastriatial nigral grafts as revealed by intracerebral dialysis. Neuroscience 22:169–178

Turski WA, Czuczwa SJ, Kleinrok Z, Turski L (1983) Cholinomimetics produce seizures and brain damage in rats. Experientia 39:1408–1411

Woodruff ML, Nonneman AJ, Baisden RH, Whittington DL (1988) Transplants of fetal hippocampal tissue impair acquisition of the Morris water maze by rats with hippocampal ablations. Soc Neurosci Abstr 14:764

Zetterström T, Brundin P, Gage FH, Sharp T, Isacson O, Dunnett SB, Ungerstedt U, Björklund A (1986) In vivo measurement of spontaneous release and metabolism of dopamine from intrastriatal nigral grafts using intracerebral dialysis. Brain Res 362:344–349

Zhou FC, Auerbach SB, Azmitia EC (1987) Stimulation of serotonergic neuronal maturation after fetal mesencephalic raphe transplantation into 5,7-DHT-lesioned hippocampus of the adult rat. Ann NY Acad Sci 495:138–152

Subject Index